P9-AGO-394

A PORTRAIT OF THE BRAIN

Men ought to know that from the brain, and from the brain only, arise our pleasures, joys, laughter and jests, as well as our sorrows, pains, griefs and tears . . . . These things that we suffer all come from the brain.

Attributed to Hippocrates, fifth century BC

ADAM ZEMAN

A PORTRAIT OF THE BRAIN

Yale University Press
New Haven and London

For information about this and other Yale University Press publications, please contact:
U.S. Office: sales.press@yale.edu www.yalebooks.com
Europe Office: sales@yaleup.co.uk www.yaleup.co.uk

Set in Minion by J&L Composition, Filey, North Yorkshire
Printed in Great Britain by St Edmundsbury Press, Bury St Edmunds

Library of Congress Cataloging-in-Publication Data

Zeman, Adam.
A portrait of the brain/Adam Zeman.
p. cm.
Includes bibliographical references and index.
ISBN 978–0–300–11416–4 (alk. paper)
1. Neurology—Popular works. 2. Brain—Popular works. I. Title.
RC351.Z46 2008
616.8—dc22

2007032187

A catalogue record for this book is available from the British Library.

10 9 8 7 6 5 4 3 2 1

For Rebecca

Don't it always seem to go that you don't know
what you've got till it's gone?
Joni Mitchell, 'Big Yellow Taxi', 1970

It is even so – Nature is nowhere accustomed more openly to display her secret mysteries than in cases where she shows traces of her workings apart from the beaten path. . . . For it has been found, in almost all things, that what they contain of useful or applicable is hardly perceived unless we are deprived of them, or they become deranged in some way.
William Harvey, Letter to Jan Vlackveld, 1657

I take apart their insides, discover the insides of their insides,
until I know the atoms of the molecules that make the cells stick.
But where is man desiring beauty?
John Martin, *The Origin of Loneliness*, 2004

To know the nature of man one must know the nature of all things.
Hippocrates

Contents

Figures

Foreword

This book introduces the brain, level by level, from atom to psyche, through tales of its disorders. It will give you fresh insights, I hope, into the kind of being that you are: physical system, living creature, conscious mind, three in one and one in three. It approaches these three aspects of our nature by way of the key elements of medicine recognised, long ago, by Hippocrates – patient, doctor and disease. The book opens a window on to each. The brain is involved in every human experience, and the book's emotional range reflects this, taking in sadness, anxiety and dismay besides desire, imagination and joy.

I am extremely grateful to the patients who have allowed me to describe their cases. I have changed circumstantial details to preserve anonymity, but have tried to retain the colour and the liveliness of the people I describe and my encounters with them.

I thank friends and colleagues, old and new, warmly, for their help. Brian Bosch, Sarah Butterfield, Stephanie Fleming, Laura Good, Sally Laird, Jon Stone, Erin Townsend, Michael Trimble and Bob Will have given helpful advice about individual chapters. Rebecca Aylward, Susan Blackmore, Micha Dewar, Sophie Zeman and two anonymous reviewers have provided detailed comments on the whole book. Matt Ridley, Bronwyn Terrill and David Salas have kindly helped with queries on specifics. Peter Tallack, my agent, helped to get things going at the outset, and has maintained a keen interest to the end. Heather McCallum, my committed editor at Yale, has worked hard to coax from me a coherent portrait of the brain. I have enjoyed working with Shelley James who is a skilled, imaginative illustrator. There will be flaws – and they are mine.

For permission to reprint poetry and prose extracts from copyright material the author and publishers gratefully acknowledge the following: Faber and Faber Ltd and Harcourt Inc. for 'Little Gidding' © T. S. Eliot 1942 and © Esmé Valerie Eliot 1970 and 'The Dry Salvages' © T. S. Eliot 1941 and © Esmé Valerie Eliot 1969, from *Four Quartets*, and *Murder in the Cathedral* in Eliot's *The Complete Poems and Plays*; Faber and Faber Ltd and Farrar, Straus and Giroux, LLC for 'Death of Orpheus' by Seamus Heaney from *After Ovid: New Metamorphoses* edited by Michael Hofmann and James Lasdun, anthology © 1994 Michael Hofmann and James Lasdun; Faber and Faber Ltd and Random House, Inc. for 'In Memory of Sigmund Freud' © 1940, 1968 by W. H. Auden from Auden's *Collected Poems*; Farrar, Straus and Giroux, LLC for 'One Art' from *The Complete Poems 1927–1979* by Elizabeth Bishop, © 1979, 1983 by Alice Helen Methfessel.

Introduction

What is this world? What asketh men to have?
Now with his love, now in his colde grave . . .
Chaucer, 'The Knight's Tale'

'What are we?' Chaucer very reasonably wondered. 'In action, how like an angel! In apprehension how like a god!' Shakespeare might have answered, always mindful that we are also a 'quintessence of dust'. Seven hundred years on, we tend to answer Chaucer's question by pointing firmly at the brain: we are what we are because we carry around in our heads this most astonishing biological device, moulding our thoughts and guiding our actions, built of around one hundred thousand million cells, making perhaps one hundred million million interconnections, the most complex system so far encountered anywhere in the universe. Surely, the brain must hold the key to human nature: understanding it will allow us to make sense of so much that puzzles us about ourselves.

If you share my excitement with this idea, and perhaps some of my doubts, read on. This book will introduce you to the brain by a route that reveals its continuity with the rest of nature, reconstructing it level by level, from atom to gene, gene to protein, protein to cell, and then through layers of increasing neuronal complexity, to the palpable substance we can – from time to time – inspect with our naked senses. The scientific discoveries which make this job of reconstruction possible are among the most remarkable intellectual achievements of the past few centuries. In the course of the ascent from atom to brain we will seek an explanation for the emergence of the brain's invisible but inseparable counterpart, the mind.

In fact, the mind will never be elbowed out of sight for long. I shall introduce each level of the brain's description through a personal story. The stories belong to people whose brains have behaved – or misbehaved – in ways that are most powerfully understood at one of these levels, from a deficiency of the atoms of a single element to the loss of some large swathe of the brain. Some of my stories are sad reminders of our fragility: 'it is terrible that such things can happen', as Karen Blixen wrote to her mother from Africa, of a speechless boy she had rescued *in extremis* from the roadside. Others tell of our great capacity to cope, adapt and grow. All speak of the human condition.

Describing the brain through the experience of people with neurological disorders will give me a chance to introduce you to my trade, neurology. Neurologists look after people with disorders of muscle, nerve and brain. We would like you to think we are scientists, and some of us are, but we really earn our daily bread by working as interpreters: we translate our patients' stories, sometimes humdrum, often extraordinary, into the scientific terms which can explain them. If we succeed, we then translate the science that makes sense of their experience into terms that make sense to our patients. Cracking the riddle posed by an illness is challenging and satisfying, not least because it may lead to a cure; getting the answer wrong, an unavoidable risk in the medical enterprise, is upsetting in equal measure. I will describe the thrill of success and the throb of failure.

It may help to know a little, at the outset, about the book's sources and plan of campaign, but if you are keen to maximise suspense, you might prefer to return to the following section after reading the rest of the book.

The opening chapter, 'I Am Tired', tells what was, for me, a chastening story, Alison's: the case of a classic neurological disorder mistaken for chronic fatigue. Alison's fatigue, and the complications that ensued, proved to have an elementary, indeed elemental, cause. The case introduces the lowest level of analysis in this book, the level of atom and molecule, the microscopic building blocks of our world – including, of course, our bodies. I introduce these microscopic constituents through the history of their discovery, as learning about the origins of ideas is often a big help in understanding them: 'the history of a science is the science'.

Chapter 2 moves from atom to gene, to explore what genes are, and how they control our brains. It takes, as an example, a rare inherited disorder of the brain that affected Charley's movement, thought and behaviour – a case that proved to be, in several senses, unique.

In Chapter 3 the focus shifts to another group of 'macromolecules' which, hand in hand with our genes, provide the chemical foundation of our lives – proteins, the 'first things' of biology. The disorder described in Chapter 3, a disorder best understood in terms of protein function, is topical – and terrible. Mercifully the threatened epidemic of this illness in the UK has not, so far, materialised.

Chapter 4 racks up the resolution of my imaginary microscope from macromolecules to the minute inhabitants of our cells, the 'organelles'. Like atoms, genes and proteins, these are found in cells throughout the body, not just in the brain. But the organelles described in this chapter are microscopic power stations that play an especially crucial role in tissues with a high energy requirement – like the brain. Dysfunction of the organelles in question, mitochondria, often causes neurological disorder. Mitochondria turn out to have an extraordinary evolutionary history, of radical metamorphosis, lending this chapter its title.

Chapter 5 introduces the brain's most celebrated resident, the nerve cell or neuron, through a case of the most common serious neurological disorder, epilepsy. The subtext of Dave's story is the discovery of the neuron, one of the major quests of nineteenth-century science.

The signals transmitted by neurons are electrical, but at the junctions between one neuron and the next the medium of the message changes, from electrical to chemical: liquid neurotransmitters float across the synapse, enabling one neuron to communicate with the next. Chapter 6 introduces this process with the help of a nineteenth-century physician, Dr Gelineau, and a twenty-first-century teenager, Lucy, who share an interest in a fascinating, under-diagnosed disorder that has recently surrendered the secret of its cause. I have allowed myself some imaginative licence here, encouraged by the importance of dreams in this condition, and by absorbing conversations with many sufferers over the past ten years.

Chapter 7 moves from the building blocks of the brain to their organisation. It examines the networks of neurons that allow us to remember, through two cases of memory disorder, those of Shona and Jed: Shona has an outlandish form of *déjà vu*, Jed a recently described form of transient amnesia.

In Chapter 8 we emerge from the microscopic scale to examine the brain as a whole. The focus is on the functions of the cortex. Jan, the hero of this tale, is in the grips of another form of memory disorder. His dementia is depleting his database of knowledge about language and the world, a database we draw on constantly to interpret our experience. In Jan's case, paradoxically, this allowed the late flowering of a neglected talent.

Patients who come to see doctors often don't have diseases. I can remember the moment that, as a medical student, I realised this with almost shocking suddenness. The medical process begins with a symptom, or several, but the match between symptom and disease is very loose. This is just as true in neurology as in other parts of medicine. Studies consistently show that around 30 per cent of patients attending neurology outpatient clinics have 'medically unexplained symptoms'. Such patients are as disabled as those whose symptoms *are* explained, and are more likely to be in some form of psychological distress. Chapter 9 examines a case like this, Jenny's – and the fascinating history of thinking about hysteria, one of the routes by which symptoms can develop in the absence of disease. Chapter 9 brings us back from the secret gardens of science into the more familiar realms of human emotion and action.

In the final chapter, 10, I recap on the journey so far – and ponder some questions that lie in the background of the preceding tales: how does a human self emerge from the 'human machine'? Can science give us a really satisfying understanding of our complex natures – or do we still need to invoke the soul to make proper sense of ourselves? The epilogue is a variation on this closing theme.

CHAPTER 1

Atom

I Am Tired

I am tired, cutting down the bracken . . .
Gaelic folk song

ALISON

All things are full of weariness. A man cannot utter it.
Ecclesiastes 1:8

My patient, Alison, was tired, and so was I. She was the seventh of my ten new cases that morning. I was standing in for a senior colleague and had assured him, with the excessive generosity of youth, that I would have no problem dealing with his work. I had meant what I said at the time but, as I read this seventh letter of referral, my already flagging spirits flagged a little further: 'Thank you for seeing this pleasant woman of 38 . . . tired for some time, especially in the mornings . . . trouble coping with the housework . . . seen by yourself a few years ago . . . Diagnosis almost certainly "chronic fatigue"' – this problem looked to have a snowball's chance in hell of a solution in the fifteen minutes left at our disposal.

What the mind does not know the eye will not see. I did my best with Alison, or so I thought, as I listened to her story and examined her weary limbs. She was indeed a 'pleasant woman' – dark-haired, dark-eyed, undemanding but in need of help. She was tired, no doubt, more so in the mornings than later in the day. She woke with a headache, which improved as the hours wore on. Slowly but surely, her symptoms were worsening. She could cope, but her fatigue had taken the joy out of her life. A look through her notes confirmed that she had seen my colleague a few years before. He had detected an abnormal blood result, further tests had not revealed a cause and matters had been left there by mutual agreement. Had I been sharper, or less stretched, I might, perhaps, have seen the light during our brief encounter. But I was young and busy: too busy to pause and think afresh about Alison's predicament, and too inexperienced for my thinking to get me anywhere much. What was certain was that I had nothing useful to say when we parted: my initial pessimism had been entirely justified. As I handed Alison a card for some blood tests and prepared to see my next patient I expected nothing more to come of our meeting, a humdrum, unedifying, 'clinical episode'. Two days later I was on holiday, purging my own fatigue, renewing my enthusiasm for life.

THE WELL OF WEARINESS

Lying down was not for Oblomov a necessity, as it is for a sick man or for a man who is sleepy; or a matter of chance as it is for a man who is tired; or a pleasure, as it is for a lazy man: it was his normal condition.

Ivan Goncharov, *Oblomov*

Oblomov, the amiable but indolent hero of Goncharov's nineteenth-century novel, is generally to be found in a dressing gown at midday.

Goncharov's introduction nicely summarises the principal causes of such a state of affairs. Until recently I lived in Presbyterian Edinburgh where working too hard and sleeping too little was the commonest cause of fatigue. Sometimes the habit of overwork develops so gradually that it goes unnoticed, as it had done for a particularly conscientious employee who consulted me on account of his tiredness. His working day ordinarily extended from 8 a.m. to 6 p.m., but over the years he had developed the habit of returning home to work a little, taking an early supper, retiring to bed at 9 p.m., and rising again at two to sneak three or four hours of paperwork during the early hours before lying in luxuriously from 5 till 7 a.m. Overwork was his normal condition, and he took much persuading that it might, conceivably, be the explanation for his growing world-weariness.

Then, as Goncharov implies, serious illness of almost any kind can be fatiguing. The cause can remain mysterious for a while, but usually a careful assessment by an experienced hand, and sometimes a few tests, reveals the underlying problem – a slowly failing organ, a deep-seated infection or, worse, an inapparent, metastasising cancer. But what of a drenching fatigue that permeates every tissue of the body, soaks the whole being in torpor, slows the system to standstill, all in the absence of any ascertainable physical disorder: is this an illness – or is it a sin?

When I fall ill, I find it hard to believe that I will ever be restored to the blessed, contented normality of health. Once I am well again, I find it just as hard to believe in the possibility of illness, that I might be stricken by one of those improbable afflictions our nature has in store. The same may possibly be true for you, and it seems to be especially hard for energetic folk to acknowledge the reality of fatigue. Industrious, striving, busy people – including of course most doctors – are simply baffled by the contradiction of chronic, disabling exhaustion in what appears otherwise to be a perfectly healthy body. It is enormously tempting to regard the problem as a moral one, a failure of will, a species of laziness, a sin – the sin of sloth.

Victims of chronic fatigue are keenly sensitive to this reaction to their plight. As a result, chronic fatigue has become one of the most politicised of ailments. Its very existence – as a medical disorder – has been debated in Parliament, and our Chief Medical Officer of Health recently found it necessary to reassert that it is a 'real and serious illness'. The heated argument often seems to turn as much on faith as on science, but science can in fact shed light on the puzzling problem of chronic fatigue.

The problem turns out to be an old one: 'neurasthenia', nervous exhaustion, a diagnosis made among women of good family in Victorian England, and the 'effort syndrome' of the trenches of the First World War are the direct ancestors of the myalgic encephalomyelitis – 'ME' – of the 1950s, yuppie flu of the 1980s and today's chronic fatigue. Are these conditions 'real'? Of course they are – common and important causes of distress and disability. Are they disorders of body or of mind? I hope to convince you, little by little, as this book unfolds, that this distinction is far more tenuous than we usually believe. But, just for the present, it is probably best to answer 'both': how could chronic fatigue and inactivity possibly fail to impinge on both fitness and feelings?

It is always dangerous to argue back from treatment to cause, but the treatments which have been shown to work for chronic fatigue are revealing: gradually increasing amounts of exercise – a 'physical' approach; and cognitive behaviour therapy – a 'talking treatment' – that aims to identify and modify the mistaken beliefs which sometimes drive our behaviour. What do these treatments aim to achieve? They aim to heal the mind via the body – and body via mind.

I assumed, on that busy morning, that Alison and I were travelling through some part of this difficult territory. I knew well enough that there are some traps for the unwary in the diagnosis of fatigue, unusual disorders which develop so inconspicuously that their signs can long be missed. I had encountered a few of these previously: Addison's disease, a lack of the adrenal hormones that we need to survive under stress, can give rise to a background of malaise and fatigue for months or years before the problem comes to light; narcolepsy, with its overwhelming need for sleep, can be mistaken for chronic fatigue, or for straightforward laziness; sarcoidosis, a mysterious tuberculosis-like inflammation can, like tuberculosis itself, cause a state of chronic exhaustion. I had such thoughts at the back of my mind. But none of them quite fitted the bill.

ATOMS AND THE VOID

In truth, there are atoms and a void.
Democritus

I returned, refreshed, from my holidays a couple of weeks later. I can't clearly recall the moment when I found out what had happened to Alison. I am surprised by this. It must reflect the protective amnesia which shrouds one's most abject moments. I think our ward round simply tailed off in Intensive Care, and as my colleagues began to recount the tale of our patient there, I realised that I knew her – that two weeks before we had been sitting face to face elsewhere in the hospital, as she told me about her fatigue. What on earth had happened?

What had happened was this. As luck would have it, Alison's visit to the hospital to see me had marked a turning point in her condition. Over the next few days, fatigue had turned into exhaustion. Her morning headaches worsened until she could hardly bear the throbbing when she woke. It was all she could do to drag herself out of bed and start the day. And then, one morning, just as my holiday ended, it was more than she could do. Her husband tried to wake her before he left for work. He stroked her face, then rocked her shoulder, then shook her bodily. Try as he might he could not wake her, for the good reason that she was no longer asleep, but in coma. The cause of her coma was elementary, or, even, elemental – betrayed, on arrival in hospital, by the blueness of her lips. When my thoughts in clinic were turning to ME and yuppie flu, they should have been directed to the absolute basics of life, the atoms and the elements that lie at the foundation of our lives – for here lay the explanation of Alison's exhaustion and her coma.

The idea that matter is composed of invisible particles, or atoms, is as ancient as the notion that a limited number of primary substances, or elements, underlies the diversity of things. The Greek philosopher Leucippus arrived at the idea of the 'atomos' in the fifth century BC by asking himself whether matter could be indefinitely divided or whether it becomes, at some point, indivisible. He decided that it must be indivisible: atoms were its smallest microscopic units. His pupil Democritus developed the theory, proposing that all atoms, 'strong in solid singleness', were identical in substance but differed in size, shape, position and speed: their combinations and recombinations gave rise to the multitudinous properties of things. These thoughts contain the seeds of modern chemistry, not to mention the germ of evolutionary theory, as these early atomists realised that those combinations of atoms best fitted to their environment would survive and prosper. The richest source of knowledge of these ancient ideas is not a textbook but a poem, the wonderful *De rerum natura*, 'On the Nature of Things', written four hundred years later, in the lucid intervals of his recurrent madness, by the Roman poet Lucretius.

These insights were prescient. They lived on, disguised and distorted, in the classical theory of the four elements – earth, air, fire and water. They encouraged the alchemists in their pursuit of the philosopher's stone – which 'will convert to perfection all imperfect bodies that it touches' – during the Middle Ages. But it was not until the seventeenth century that these ideas began to take their modern form. The troubled, deeply religious Oxford scientist Robert Boyle recognised that there were not four but numerous primary substances, all composed of particles. He realised that these substances, the elements, could exist in solid, liquid and gaseous forms. In *The Sceptical Chymist*, published in 1661, he defined the elements as 'certain primitive and simple, or perfectly unmingled bodies; which not being made of any other bodies, or of one another, are the ingredients of which . . . perfectly mixed bodies are immediately compounded, and into which they are ultimately resolved'. One hundred years later the Parisian scientist and administrator Antoine Lavoisier made it his overarching aim to identify these 'undecomposable, elementary substances' which, through myriad combinations and transformations, underlie the shifting appearances of the chemical world. In 1789, the year of the French Revolution, Lavoisier proposed 33 candidates for this role, of which 25 have stood the test of time.

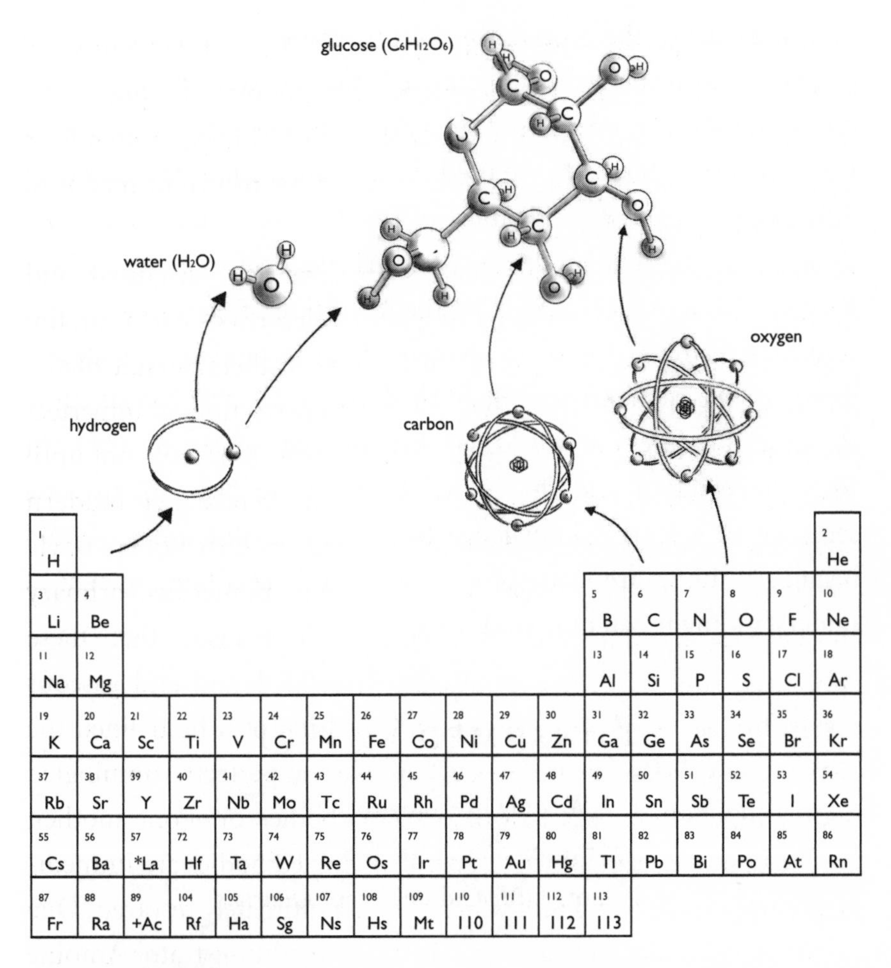

1. The Periodic Table
The periodic table shows the elements, organised according to their chemical properties. Some heavier elements are omitted: 117 elements have been observed. All the elements that abound in living things lie in the first four rows: these include hydrogen (H), carbon (C), nitrogen (N), oxygen (O), sodium (Na), phosphorus (P), chlorine (Cl), potassium (K), calcium (Ca), iron (Fe). The atoms of elements combine to form molecules of compounds: the figure shows the very simple molecule of water (two atoms of hydrogen and one of oxygen) and the more complex molecule of glucose (six atoms of carbon, six of oxygen, 12 of hydrogen).

At the start of the nineteenth century, the son of a Westmorland handloom weaver, a devout, reclusive Quaker, John Dalton, reintroduced the term 'atom' for the smallest ingredients of the elements. Just as Democritus had proposed, he recognised that the atoms of different elements were identical in substance but differed in size. By the time the Russian chemist Dmitri Mendeleyev published his Periodic Table in 1869, clarifying the identities, properties and relationships of the eighty or so elements then recognised, the atomic theory of the elements was established. It is absolutely fundamental to understanding the world we live in: the American physicist Richard Feynman said that if we were able to pass on just one sentence of scientific knowledge to another civilisation it should begin: 'All things are made of atoms . . .' . It is also fundamental to understanding much human disease.

CONSUMED BY FIRE

We only live, only suspire,
Consumed by either fire or fire
T.S. Eliot, 'Little Gidding', *Four Quartets*

Alison's predicament was a case in point, as it turned out. What had happened, as I discovered during that dismal ward round, was simple enough. Her breathing had gradually failed overnight. Her blood had been starved of oxygen and had become overladen with the main waste product of our body's chemical processes, carbon dioxide. The brain has an unquenchable thirst for oxygen: in its absence consciousness rapidly fails in less than a minute. But matters, mercifully, had not progressed too far. Once diagnosed, the disturbance of the blood gases could be reversed by a short spell of artificial ventilation, which rapidly restored both chemistry and consciousness. It cured her headaches, which had signalled the accumulation of carbon dioxide in her blood as she slept, and it began to renew her *joie de vivre.* The cause of this near-catastrophe soon came to light: a profound, selective, weakness of the muscles which open the lungs to life-giving air, a nearly fatal faltering of the bellows which fan the fires of our existence.

The notion that the air contains an active principle that is consumed both by flames and by living creatures, both by inanimate combustion and by animate metabolism, dates back to the seventeenth century at least. It was well known by then that an animal enclosed in a glass vessel would die after some minutes, and that it would do so more quickly if a lighted candle were placed in the vessel too: 'it clearly appears that animals exhaust the air of certain vital particles . . . that some constituent of the air absolutely necessary to life enters the blood in the act of breathing'. Richard Lower, who published his *Tractatus de corde* in 1669, had discovered that this 'spiritus nitro-aereus' changed the colour of the blood as it passed through the lungs, from the dark purple which runs in our veins to the bright red of our arteries. His colleague Robert Hooke, Boyle's assistant in Oxford, showed that the act of breathing itself was not essential for this transformation, provided a current of air was continuously blown over the lungs.

The precise nature of the active principle eluded discovery for a hundred years. Heating mercuric oxide in 1772, the Swedish pharmacist-chemist, Karl Scheele, isolated 'fire air', the fraction of the atmosphere which, he found, supported fire. Scheele was a modest but intrepid man, given to sampling the properties of novel chemicals at first hand: he died in his forties, probably from poisoning by mercury. His *Experiments on Air and Fire* had not yet been published when, in 1774, using the same method, the English chemist Joseph Priestley also obtained a novel 'air' in which 'a candle burned with an amazing strength of flame'. On a trip to France, Priestley shared his discovery with Antoine Lavoisier. Within a few years Lavoisier had named the substance 'oxygen' and clarified its role in fire and life. The bright flame of the candle and the warm glow of the body are indeed alike in their chemical nature. Each depends on 'oxidation', a chemical bonding with oxygen, drawing a measurable weight of the gas from the surrounding air. We are all exquisitely regulated, very slow burning, chemical stoves, 'one of those fires without light'.

Lavoisier went to the guillotine in 1789, condemned because the 'Republic has no need of savants'. By the time of his death, it was clear that four of the elements he had identified – hydrogen and oxygen, both of which Lavoisier named, together with carbon and nitrogen – were the prime ingredients of living things. We depend absolutely on the oxygen of the atmosphere to keep the flame of our metabolism alight: it is the gas in which we burn the fuels that sustain us – and our brains. But our bodies also abound with 'fixed' oxygen, which constitutes 65 per cent of our body weight: combined with hydrogen in water, and with hydrogen, carbon, nitrogen and phosphorus in the larger molecules from which our bodies are constructed – like DNA, carbohydrates, fats and proteins. We accommodate a few other elements, to be sure: a watery solution of sodium, chlorine, potassium, magnesium and calcium, the elements dissolved in the seas in which all life was born, still bathes our tissues, and our evolution has put traces of heavier elements, like copper, cobalt, zinc and iron, to ingenious uses in the nooks and crannies of our cells. But nonetheless the elements that compose us can just about be numbered on our fingers and toes.

The atoms of each element are its smallest distinctive units. They are very tiny indeed: the size of a hydrogen atom is around a ten-millionth of a millimetre. Atoms combine with one another to form molecules, gangs of mutually attracted atoms. In this way, elements team up to form compounds, like the carbon dioxide we exhale, with one atom of carbon and two of oxygen in each of its molecules, CO_2, and the water we drink, H_2O. Are atoms really the smallest divisible units of matter? Well, alas for simplicity, of course they are not: much of the twentieth century's physics was devoted to exploring the world within the atom, its central nucleus and orbiting electrons, and to discovering the vast quantities of energy that can be derived from splitting or fusing these nuclei. The quest for the ultimate constituents of matter continues today, along the path Leucippus and Democritus first opened up, invoking ever more abstruse entities, such as leptons, quarks and strings. But at the level at which we mostly lead our lives, using naked eye and bare hand, the inventory of the elements, chemically pure, atomically uniform, like gold and silver, lead and iron, carbon and oxygen, is fundamental. Though oxygen is the particular hero of this chapter, carbon deserves a special mention. Its atoms have a unique propensity to enter into stable combinations with one another, and with the atoms of other elements. Carbon contributes to nearly ten million known compounds: the chemistry of carbon, 'organic chemistry', is therefore the chemistry of life.

We have borrowed these elements, the stuff of our lives, from the stars. The atoms from which we are built were formed there, by nuclear fusion, at temperatures so astounding that the simplest nuclei, of hydrogen and helium, fused together to generate their heavier companions, later to be flung around the universe as dying stars erupt. We really should be more startled by the exotic nature of our basic components. We think of physics as the science of matter, biology as the science of life, and tend to look to biology for an explanation of our human selves. But whatever the ultimate constituents of matter turn out to be, you and I are as intimate with them as it is possible to be: we are built from them – we are they, no more, no less.

ELEMENTAL AFFLICTIONS

We live at the bottom of an ocean of the element air
Evangelista Torricelli

You may have sampled symptoms like Alison's if you have climbed to altitudes at which the falling oxygen pressure in the thinning atmosphere begins to take its toll – typically over 10,000 feet. A sense of mounting effort and worsening breathlessness may be followed by full-blown altitude sickness, ushered in by nausea, dizziness, insomnia and a pounding headache. The sudden decompression of an aircraft at 30,000 feet results in more dramatic oxygen deficiency: unless the supply is swiftly restored, the 'time of useful consciousness' before the brain fails is a mere minute.

It is unusual for neurological diagnoses to focus on a single element, but there are some important exceptions. Sodium, the metallic element contained in sodium chloride, our table salt, and calcium, familiar from calcium carbonate, chalk, are both abundant in our body fluids, and essential for electrical signalling in nerves. If calcium levels fall in the blood – as they will do if the hormone which maintains them, parathyroid hormone, is underproduced – nerves become hyper-excitable, entering a state of 'tetany': tingling, muscular spasms, often affecting the hands and feet, and even convulsions ensue. Hyperventilation in the course of panic attacks, a much more common occurrence than deficiency of parathyroid hormone, has a similar effect: the resulting reduction in the level of carbon dioxide in the bloodstream makes it more alkaline than usual, which in turn reduces the amount of electrically active calcium. By contrast, if parathyroid hormone is overproduced and levels of calcium rise, our muscles become leaden, action effortful and the mind befuddled. Abnormal levels of sodium in the blood, whether low or high, also disturb the workings of the brain, to cause confusion, seizures and finally coma. These are generally reversible once the cause is spotted, but over-rapid correction of depressed sodium levels can cause irreparable brain damage.

Heavier elements sometimes give rise to atomic disease, most commonly when we are poisoned by them. Hippocrates, in 370 BC, suspected that a patient's abdominal pain was due to his work as an extractor of metallic lead from ore, the first recorded case of 'Saturnism', named after the alchemical term for lead. Daniel Defoe, in the eighteenth century, described an encounter with a miner of Derbyshire lead who seemed to have risen from 'the dark regions' themselves, 'pale as a dead corpse, his hair and beard a deep black, his flesh lank, and, as we thought, something of the colour of lead itself'. The metal finds its way into the nervous system, damaging our nerves and the blood vessels of the brain, but lead poisoning, common in the days of lead piping and lead-based paints, is nowadays a rarity.

Lead was never intended to cohabit in our cells, but some heavy elements are used by the body in trace amounts for specific, highly regulated, purposes. Iron is one such, helping to transport oxygen in our blood cells and to bind it elsewhere; copper another. These guests need careful handling by their hosts: the consequences can be dire if these elements are allowed to accumulate in the wrong places or fail to get where they are needed. The body's failure to handle copper normally underlies two grave inherited diseases. Both are thought to result from inherited abnormalities in proteins that 'transport' copper to the sites where it is needed. In Wilson's disease, copper accumulates in eye and brain (as well as liver), causing disorders of movement, mood and mind which are sometimes mistaken for signs of a primarily mental disease before the condition reveals itself fully. It usually comes to light in teenagers or in early adulthood. It can be effectively treated with chemical remedies that leach the excess copper from the body. The related, but more severe, condition of Menke's disease causes poor growth, lax skin, kinky hair and neurological decline in infants. In this condition the body fails to incorporate copper in the small group of enzymes – proteins that facilitate chemical reactions in our cells – in which it plays a vital part. These include the enzyme cytochrome oxidase, which returns us to the atomic hero of the chapter, as cytochrome oxidase completes the process by which our cells generate energy in the presence of oxygen. If copper is important in normal functioning, one would predict a syndrome of copper deficiency. New disorders are constantly being recognised, like new species: in 2001 copper deficiency was recognised for the first time as a cause of neurological illness. It causes difficulty in walking and disturbed sensation in the legs as a result of dysfunction of the spinal cord, usually in people who have been absorbing copper poorly following stomach surgery, often performed many years before.

Conditions like Wilson's disease, and the heavy elements which cause them, are biological rarities. The principal elements of life, hydrogen, carbon, nitrogen and oxygen, are lightweights in the periodic table, with atomic numbers, reflecting their atomic weights, of

1, 6, 7 and 8. The key to their role in our lives is their fertile capacity to combine with each other, forming the molecules of life, our genes, proteins, sugars and fats. These are the constituents of all our organs, including the brain. It is easy to forget that the myriad possibilities of human life flow ultimately from the humble but mysterious atoms that compose us.

LIFE-FORGIVEN

Life-forgiven and more humble,
able to approach the Future as a friend . . .
W.H. Auden, 'In Memory of Sigmund Freud'

Alison's story ends more happily than I feared it might. Life forgave us both. Her underlying condition proved to be rare indeed, so rare that we reported it in a scientific paper as a hitherto unreported consequence of an uncommon muscle disorder, 'multicore myopathy' – named after the cylindrical cores that are visible under the microscope within the muscle fibres. Alison's myopathy especially affected her diaphragm, the two-humped muscle which lies beneath the lungs, and flattens, pushing out the abdomen, when we inhale. This was the reason it had escaped our notice – routine neurological examination fails to assess the diaphragm.

This muscle is particularly important to us in sleep, as gravity no longer keeps our abdominal organs in their place when we lie down. If the diaphragm fails, breathing in sleep can fail too, as Alison's did. But over the past half-century respiratory physicians have refined simple techniques for assisting breathing overnight, maintaining the pressure of oxygen in the lungs. One of these suited Alison well, restoring her oxygen levels to normal, extinguishing her headaches and greatly relieving her fatigue. Her muscles have gradually weakened elsewhere, and this we could not cure, but her life was no longer at risk.

As for me, I felt foolish, though no one rubbed salt in the wound. I wanted to hide. But on second thoughts, instead of nervously concealing them, we should examine, even celebrate, our failures and mistakes. Rather than being negligent or shameful, as a rule, they are a fact of life, a plain reflection of human imperfection. They have so much to tell us about ourselves that they deserve our scrutiny, offering our best chance to learn and grow. With their help we can avoid repeating history. Such pious thoughts, needless to say, were far from my mind that day in Intensive Care. In fact, despite my holiday, I suddenly felt very tired.

CHAPTER 2

Gene

Don't Fidget

THE CLASSIFICATION OF FIDGETS

You and I are built from atoms, nothing more and nothing less. But over the first thousand million years or so of the earth's life, atoms began to find their way into the wonderfully fertile combinations that we recognise as *living*. Indeed, the similarities that unite all forms of life on earth, from the broad bean to the brontosaurus, are more profound than the eye-catching differences between them. This chapter will explore the explanation for this fundamental bond.

Despite their underlying unity, living things can be divided into two great classes on the basis of a single key feature – their ability, or inability, to get up and go. By and large, animals can and plants can't. Most applications of this criterion – say to cheetah versus cedar or octopus versus oak – are straightforward, though as usual in biology there are some problematic borderline cases, like bounding bindweed versus the lazy limpet. We humans fall, of course, well over to the footloose side of the line, and large parts of our brains are devoted to detecting, assessing, plotting, preparing and executing movement. Much of a neurologist's life, therefore, is concerned with the diagnosis of disorders of movement, including a miscellany of fidgets.

Many of these are perfectly normal superfluities of movement, although they occasionally cause concern, particularly to parents. You probably indulge in a few of them yourself, drumming your fingers on the table while you sit in a dull meeting, picking at the rough edge of a nail with your fingers or your teeth, clearing your throat, tapping your foot or just whistling an oft-repeated tune while you wander. These innocent motor habits are half-conscious outflows of 'nervous energy', reminders that our systems are geared up and ready for action when little – perhaps too little – is expected of them. The wriggliness such habits reveal is an abiding character trait: we all have more or less restless motor personalities. Try reading aloud to a child on a sofa. The wriggly kind make this activity remarkably hazardous, dislodging errant cups and bits of food from the sofa's arm as they rotate their elbows, hips and knees before driving them hard into the reader, while their more comfortable friends and relations quietly relax into the contours of the chair and fall asleep.

Some of us, most often children, elaborate these commonplace habits into embarrassing, unignorable sniffs, coughs, blinks and fidgets. These are known in the trade as 'tics', mannerisms that can be suppressed with an effort but that tend to break out again when the effort is relaxed. Sufferers from Gilles de La Tourette's syndrome, which lies at one extreme of the spectrum of restlessness, feel compelled to produce a multiplicity of sudden tics, including grunts and obscenities, which startle passers-by in public places. Their jerks and cries fall in the border zone between intended action and involuntary motion. The 'Touretter' is the victim of urges he would rather not have, yet he experiences the urges as his own. Other excesses of movement are more clearly experienced as imposed from without, further removed from any possibility of voluntary control – like the disconcerting 'hypnic jerks' of the whole body which can startle us for a moment as we drop off to sleep, or the tremor which rattles our cup in its saucer during an anxious encounter.

But the focus of this story is on yet another group of involuntary movements. These play restlessly over the face, limbs and trunk; if quick and jerky they are named after the Greek word for dance, 'chorea'; if slow and writhing, they go by the name 'athetosis'. As they often contain a mixture of the two types they are known generically as 'choreoathetosis'. Sufferers from choreoathetosis may or may not be concerned by their ceaseless movements: but either way it is more than they can manage to keep still for a matter of seconds. Whereas many of our twitches and jerks can be caused by mishaps in several parts of our nervous system, choreoathetosis is a reliable pointer to trouble in a particular set of brain regions, buried deep beneath the wrinkled cerebral cortex and somewhat inelegantly named the 'basal ganglia' (sketched in the figure in the glossary on page 215).

FIRST MEMORY

Chorea framed Charley's earliest memory. He and his mother were standing in the kitchen on a cold winter's morning in Wales. He was six years old. Both he and his mother were about to be late – he for school, his mother for work, at the colliery where she cooked for the miners. She was brushing his hair, or trying to, but as children are prone to, and Charley much more so than most, he evaded the brush, ducking, leaping, skipping away from the bristles. All of a sudden his mother's patience broke: she let out an exasperated shriek and Charley felt the bristles biting into his thigh, bare at the edge of his short school trousers, as she redirected her aim from his head to his tail. The hairbrush, a precious one, broke. That was all:

the memory was fragmentary, as our first memories tend to be, but for some reason the recollection of the sudden pain, his mother's uncharacteristic rage and the broken brush had lived in his mind for fifty years.

Charley's teachers, also, were sometimes exasperated by this otherwise amiable child's inability to stay at his desk or stand in line for more than a few seconds without drumming a tune on the wood or dancing a jig. It was natural that when the time came to leave school Charley chose a career which would keep him up on his feet and in motion: he trained as a chef. He was an arresting figure as he worked: tall, dark, powerfully built, slightly shambling, never still. Although he worked hard and well, a jerking, clumsy quality in his movements explained the occasional spills and burns which stung almost as keenly as his childhood punishment.

Marion met him in his late twenties. She was immediately attracted to this warm and bearish man. His undeniable eccentricity, which had been offputting to some other women, only increased her interest in him. Affection grew on both sides and they married a few months later. Marion's fondness for Charley did not falter over the years to come although as time passed, and his restlessness worsened, he became an increasingly odd and awkward companion.

KITH AND KIN

The key to the kinship of all living things, broad bean, brontosaurus, you, me and Charley was fashioned about three thousand million years ago, three-quarters of the age of the earth, when atoms of several of the lighter elements, hydrogen, oxygen, nitrogen and carbon combined under the fertile chemical conditions of the time to form a molecule with a remarkable new property: the ability to get other such atoms to produce copies of itself. At the outset many different self-replicating molecules may have competed with one another for survival and success, but all currently existing forms of life on earth depend on a single class of biochemicals to perpetuate their kind down the generations – nucleic acids, which are the building blocks of DNA, the chemical substance of our genes.

The ability of nucleic acids to reproduce themselves lies at the heart of life. But in the highly developed state of living things today, nucleic acids act from a distance, through intermediaries: they shape our development and our day to day lives by spelling out the structure of the thousands of proteins which compose our being, using a four-letter alphabet, A, C, G, T. These letters stand for the 'nucleotide bases', adenine, cytosine, guanine and thymine which are strung like beads on the rungs of a ladder along the molecule of DNA (see Figure 1, p. 13).

Your DNA and mine, therefore, has two main functions. The first is to act as a library of genetic knowledge that can be handed down from one generation to the next. This is possible because a strand of DNA, with its string of nucleotide beads, can be used as a template for the manufacture of another, complementary strand: adenine pairing with thymine, guanine with cytosine. The 'double helix' of the DNA molecule consists of two such complementary strands. When the time comes to 'replicate', the strands unzip, and each can be used to instruct the formation of a complementary strand, each molecule thereby mothering two identical offspring.

The second main function is to give instructions for building proteins. A series of three of the bases, a 'base triplet', specifies that a particular amino acid – the building block of a protein – should be inserted into the corresponding protein. A series of triplets, comprising a single gene, specifies the order of amino acids in the

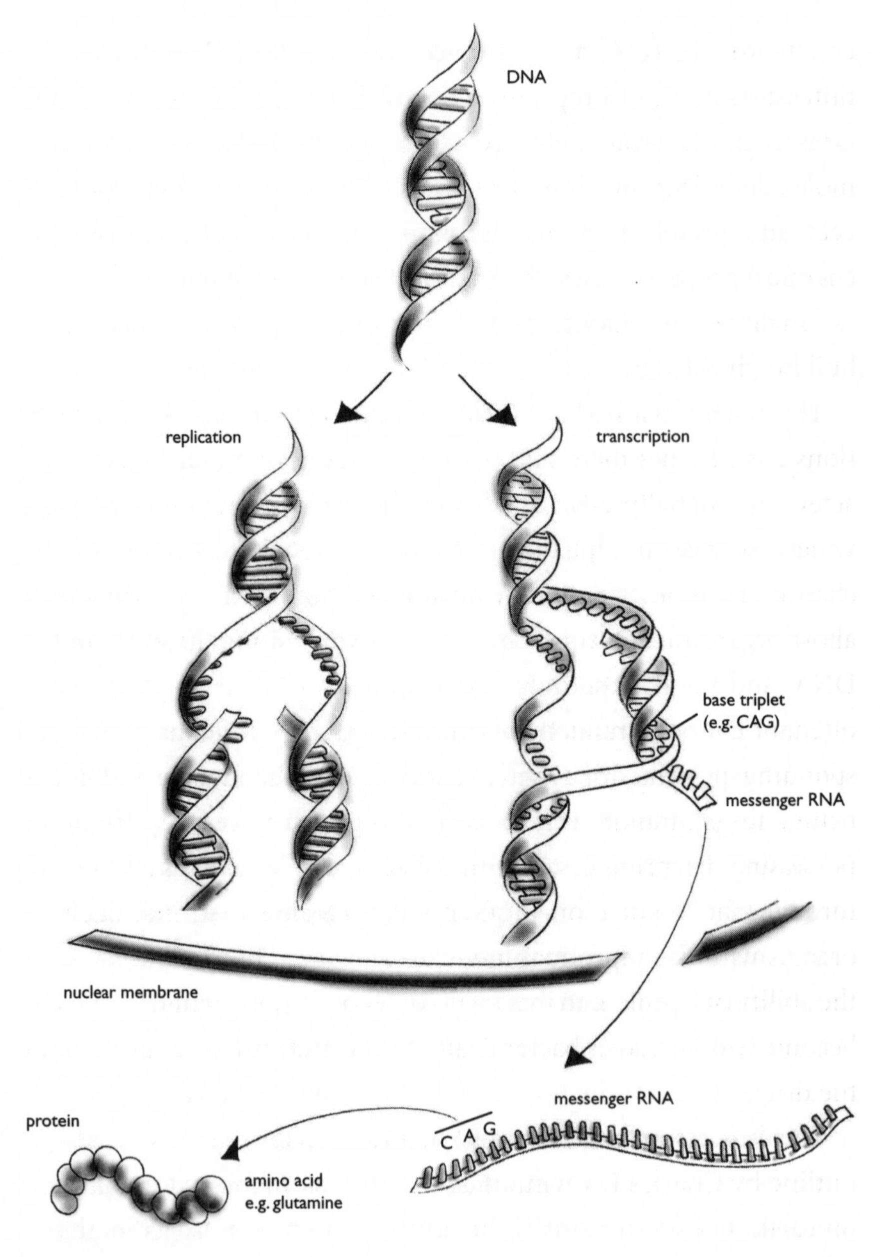

2. DNA
The DNA molecule is a double helix, in which two complementary strands wrap round each other. To replicate, the two strands come apart, so that each can be used as the template for the formation of a second complementary strand. DNA controls the production of proteins by sending a complementary strand of RNA (messenger RNA) out of the nucleus, into the outlying cell: there the base triplets in the messenger RNA (for example CAG) specify the selection of a particular amino acid (for example glutamine) which takes its place in the corresponding protein.

entire protein. 'CAG' for example codes for the amino acid glutamine: a succession of repeated CAG triplets lying within a gene indicates that the protein it specifies contains a series of glutamine molecules joined in a line. This matters because the order of amino acids in a protein molecule determines its future role: its shape, its chemical properties and its functions in the body – and proteins, as we shall see in Chapter 3, are the body's builders as well as its building blocks.

The fidelity with which DNA reproduces itself between generations ensures that the varieties of living things maintain their character. Occasionally DNA undergoes haphazard 'mutations' – for a variety of reasons, including damage to DNA by radiation and chance errors in the process by which DNA is copied. Mutations allow organisms to evolve, as once in a while a mutation alters the DNA, and the corresponding protein, advantageously. Much more often, of course, a random disturbance of DNA will alter the corresponding protein disadvantageously, but, *if* the change is for the better, the mutation will increase the chances that the creature possessing it reproduces, and the altered gene will gradually spread through the population. This can be studied experimentally in organisms which reproduce fast, like bacteria. A mutation enhancing the ability of a protein to inactivate antibiotics, for example, will soon become widespread as bacteria lacking the mutated gene succumb to the drug.

This process, of mutation and natural selection, first proposed in outline by Charles Darwin, underlies the astonishing variety of life on earth. It also accounts for its unity, as all living things share the life-conferring biochemistry of DNA. Indeed, we share more than simply DNA itself: we share our genes. There is (very) roughly a 30 per cent overlap between the DNA of mankind and flowering plants, a 40 per cent overlap with the fruit fly, 90 per cent with the mouse, 99 per cent with the chimpanzee.

In 2002 scientists undertaking the 'Human Genome Project' completed the mapping of the entire sequence of base triplets throughout the 23 chromosomes, the giant molecules into which our DNA is packed. Their work has revealed that the recipe for building and maintaining the human form requires about 30,000 genes, rather fewer than previously believed. The mapping is a huge achievement, but it is easy to misunderstand the nature of the resulting chart.

Imagine creating a map of the surface of an alien world from space, a two-dimensional description of its salient surface features. Imagine then arriving at the surface to discover that while the map is accurate, the terrain behaves in completely unexpected ways, transforming itself around you, producing a marvellous and apparently unpredictable series of forms – spirals, temples, citadels, cathedrals – which emerge from the landscape and merge back into it, only to metamorphose into other equally astonishing creations. Contemporary biologists are in a similar predicament to these extra-terrestrial explorers: equipped with a carefully drawn two-dimensional map, but decades or more from a full understanding of how the terrain it describes produces the almost infinite variety of living forms. Take, for example, the case of a recently discovered 'language gene', FOXP2, mutated in a family with a severe disorder of speech. FOXP2 turns out to belong to a family of 'forkhead genes', coding for proteins that regulate the activity of various other genes using a structure evocatively named the 'winged helix': the biological links between damage to FOXP2 and a disorder of language undoubtedly exist, but will need an immense amount of unravelling.

A SNAPPER-UP OF UNCONSIDERED TRIFLES

'My father named me Autolycus; who being, as I am, littered under Mercury, was likewise a snapper-up of unconsidered trifles.'
Shakespeare, *A Winter's Tale*, IV, iii, ll.24–6

Charley held down his job and his marriage for twenty years. But as time passed his restlessness seemed to spread from his limbs into his soul. Work became too confining and he quit. Unoccupied during the day, Charley had to be on the move: he felt compelled to stalk the streets, ambling endlessly through parts of town he would not normally visit. As he walked, he gathered – anything, everything, pieces of wood, old bottles, empty cans, fragments of glass and smooth pebbles. They began to fill the garden shed and, once that was packed, to fill the house – discarded debris which Charley insisted might some time come in useful, though how or why was never clear.

Despite Marion's growing exasperation, and her shrinking living space, he refused to part with them. She was troubled by a childish quality in this hoarding, a paradoxically aimless stubbornness. She wondered whether his haphazard collection might be Charley's attempt to shore his life up against a sense of impending chaos. She was also aware that a childish quality had crept into his relationship with her. Like a three-year-old, he was unable to let her be: Charley and his rambling conversation would follow her around, into the garden, into the toilet. There was no longer anywhere to hide.

Shortly before I met her she had been startled to catch a glimpse through a window of Charley setting off on one of his expeditions, huge, bearded, lumbering, unkempt, lost in conversation with himself: she saw him for a moment as she imagined others must see him, and the sight shocked her. A while back she might have tried to share the picture with her husband. Charley used to be able to cast a humorous eye over his own idiosyncrasies. Now he was irritated or simply indifferent if she pointed them out. And she had to acknowledge that they were becoming more outrageous. Despite her remonstrations, he had taken to marching naked through the garden to rummage in his shed. She had never thought she would be scared by Charley, but perhaps the time was coming when she ought to be.

Charley had seen several doctors by the time that Marion began to step back a little from their marriage – a GP or two and a psychiatrist before coming the way of a neurologist. It was easy enough for his doctors to list the elements of his predicament – a loss of the normal social inhibitions, a compulsion to hoard random possessions, and a disorder of movement, sometimes fidgeting, sometimes writhing or lurching, a form of choreoathetosis – but much more difficult to come up with a good explanation for this lifelong, slowly worsening condition of inner and outer, mental and physical, restlessness.

THORN CELLS

Acantha: Greek, a thorn

Charley's trips to my clinic were like visits from a good-natured whirlwind. The consulting room, normally perfectly well able to accommodate four or five people, seemed too small for Charley all on his own. He would bob around in it for a minute or two before perching briefly on a chair: he never exactly sat down, and I never felt entirely in control. There was a sense that Charley came to the clinic out of kindness to me, a force of nature bending itself to answer my questions for as long as its powers could be contained –

and then he was gone. If I was quick-witted I might manage to slip in a few extra enquiries while I had Charley pinned down on the end of a needle to draw a blood sample. In a sense, of course, Charley was absolutely right: there was just a chance that I would find out what was wrong, which interested him rather little, but virtually none that I could cure him, which interested him a great deal.

The standard approach to diagnosis in neurology is first to establish which part of the nervous system is in trouble, *where* the problem lies, then to ask why: *what kind* of pathological process might be at work. The first question was relatively easy in Charley's case: as I mentioned, chorea is a reliable pointer to trouble in the basal ganglia, a group of cell clusters deep in the brain which help to organise our movements. The stiffness, slowness and tremor of Parkinson's disease, for instance, are caused by a lack of a particular neurochemical, dopamine, in this part of the brain: chorea represents the opposite extreme of its activity. The report of Charley's brain scan in his notes said it was normal. But even a well-trained eye can miss subtle changes of appearance unless directed specifically to the point of interest, especially if the changes are symmetrical. When I looked closely with a radiologist colleague we agreed that in fact Charley's basal ganglia had lost much of their substance. The heads of the caudate nuclei are key components of the basal ganglia. They are normally visible as plump convexities at the margin of the ventricles, the fluid-filled spaces at the centre of the brain. Charley's had shrunk down to almost indiscernible streaks.

But Charley's chorea was only a part of the story. What about his eccentricities of behaviour and of personality? Could these also be linked to changes occurring in the basal ganglia? They probably could. We tend to draw sharp distinctions in our thinking between our movements, thought and behaviour but the brain does not always respect these. The basal ganglia play a role in all three. Fascinating recent evidence from patients undergoing surgical treatment for Parkinson's disease shows that direct electrical stimulation of the basal ganglia can switch on and off dramatic changes in mood, from depression to mania, with corresponding changes in behaviour – not unlike the changes that had much more gradually overtaken Charley. From what we know of their roles to date, it is entirely plausible that altered function in the basal ganglia could simultaneously give rise to fidgets of chorea, loss of social inhibitions and compulsive, ritualistic behaviour, all the main features of Charley's case.

If his lifelong and still developing affliction could be traced back to some trouble occurring in these regions deep in the brain, why had it struck and what was it? The first condition any neurologist would think of in a patient with chorea and a changing personality was described by a New England physician, George Huntington, at the end of the nineteenth century. Huntington's chorea is the most common grave inherited neurological disorder. It usually begins in adult life, often a few years after sufferers have had children of their own and transmitted the disorder to another generation. It causes chorea, progressive disorganisation of thought, and alteration of mood largely due to a gradual withering away of parts of the basal ganglia. It shortens life, but the physical and intellectual decline occurs very slowly, sometimes painfully slowly, over one to two decades. The condition has a peculiar characteristic, known as 'anticipation' – as it descends through the generations it tends to occur earlier and more severely. This results from its genetic basis: the gene which causes Huntington's contains a 'base triplet repeat' of the kind I described earlier: the bases CAG recur in series within the gene which codes for a protein called 'Huntingtin'. This means that Huntingtin – yours and mine, for this protein is present in all of our brains – ordinarily contains a linked series of glutamine building blocks. Trouble starts if the number of CAG repeats within the gene crosses a critical threshold, around 30. When it does so the triplet repeat becomes unstable, increasing in length with the passage of each generation, giving rise to an increasingly aggressive disease. The altered protein accumulates within cells in affected brain regions: it is not yet clear how accumulation relates to the death of brain cells in Huntington's disease.

But Huntington's disease was never a likely explanation for Charley's lifelong disorder. Matters would have moved faster had this lain at the root of his problems. Although I made sure he did not have Huntington's by asking the laboratory to look for the relevant stretch of DNA extracted from a sample of his blood, I clearly needed to chase the rarer, more exotic causes of chorea. As I browsed through Charley's notes, and the results of previous tests, I noticed some references to 'burr cells'. These are abnormal cells in the blood which look spiky under the microscope, like the burrs you find on your sweater at the end of a walk through the woods. The blood cells in question are the red blood cells, the kind which carry oxygen from the lungs around the body. They are usually smoothly rounded cells, shaped like flattish doughnuts which have never quite managed to achieve the hole in the middle – and they should be totally spikeless. I knew of no connection between burr cells and chorea. But I wondered whether 'burr cells' might have been confused with another kind of misshapen red blood cell which I had fleetingly encountered as a neurologist – acanthocytes. These turn out to be named after a more piercing hazard in the woods, the thorn. Common sense suggested that thorny acanthocytes might be confused with spiky burr cells. A fresh blood sample, a conversation with a helpful haematologist, and Charley's peculiar symptoms began to fall into place.

OF GENES AND GYRATIONS

Of the 30,000 or so genes contained in each of our cells, more are switched on in the brain than in any other organ. This is to be expected, as the brain is by a very long way the most complex entity we have yet encountered in the universe. Each of your genes is present in two copies, one from your mother and one from your father. A gene determining a particular characteristic, like brown eyes, may be 'dominant', that is to say it tends to exert its effect regardless of its opposite number, or 'recessive', with effects which go unnoticed unless the genes from both mother and father convey the same instruction, for instance to grow red hair. The existence of recessive genes helps to explain how inherited characteristics, including inherited diseases, can appear abruptly, without any previous family history of similar events.

These examples of genetic determination, brown eyes and red hair, sound straightforward enough – but remember the metaphor of the map drawn from space: the path from a gene to its manifestations in any organism, including us, the path from 'genotype' to 'phenotype', is usually complex, incompletely understood and contingent: complex because the route from gene to a visible characteristic involves numerous intricate steps; incompletely understood because we have so much more to learn; contingent because a great many influences, including both other genes and features of the environment, can sway the expression of a gene along the way.

Take Charley's thorn cells, for example. These cells, acanthocytes, are a marker for a small family of rare disorders causing both a change in the appearance of blood under the microscope *and* the loss of certain nerve cells in the brain and outside it, in the nerves that run to muscles from the spinal cord. A disorder in this small family,

McLeod's syndrome, turned out to be the explanation for Charley's unusual story: it can give rise to a slowly worsening restlessness – chorea – accompanied by a gradual coarsening of personality and decline of intellect.

What do thorny red cells have to do with chorea or the brain? The link is indirect. The help of colleagues with expertise in the genetics laboratory allowed us to pinpoint the molecular cause of Charley's illness. It resulted from the deletion of a single molecule, a single nucleotide 'base', from a gene that normally enables red blood cells to manufacture a protein anchoring other proteins in the red cell's wall. The loss of this single base made nonsense of the genetic message, fouling up the production of the protein. Lacking this anchor the red cell wall became unstable, giving rise eventually to the thorny deformity seen in Charley's blood film. The same gene is thought to be active in the brain: for reasons we do not yet know, lack of the corresponding protein there shortens the lives of cells in the basal ganglia required to keep our movements and our minds on track.

Confronted with a patient like Charley, isn't it possible nowadays simply to turn to the human gene map and look up the relevant gene? Often not. The map is a great help in navigating around our DNA, but it neither labels the functions of genes nor reveals the layers of cause and effect between a disordered gene and a human disease. All the same, genetic research on disordered movement – like most of the rest of human disease – is booming. The identification of the gene causing Huntington's disease in 1983 was a landmark. Most of the genes that cause acanthocytosis have been lassoed in the past ten years. Around 60 genes causing inherited forms of incoordination and unsteadiness and a dozen causing inherited Parkinson's disease have been described, at a rapidly accelerating pace; the genetic changes that lie behind the cries and jerks of Tourette's syndrome should be located soon.

Delivering really effective treatment for these disorders, based on our new knowledge of their genetic basis, is, so far, an unrealised ambition. For Charley, the discovery of the cause of his predicament provided cold comfort. I was unable to rescue him from his fate. His last few years were spent in psychiatric custody – there was plenty of evidence that he had become a serious risk to himself and to others, though he never accepted this. He died of heart failure, one of the complications of his illness. He donated his brain for research.

The path between our genes, our bodies and our behaviour is an exceedingly tortuous one, but there is no doubt that it exists, and its map-makers are hard at work. The discoveries by geneticists over the past hundred years have explained once and for all how matter can give rise to life. They have revealed the underlying unity of life in the fertile powers of DNA. These discoveries are increasingly explaining how disorders like Charley's result from molecular changes in our genes. These changes, in their turn, disrupt the manufacture of the proteins by which, and from which, we are built. Such discoveries may in the future allow us to cure, or at least ameliorate, Charley's rare syndrome and others like it – but not yet. When next you fidget, or the child on your lap wriggles his knee into your groin, spare a thought for the family of genes that shapes our movements.

CHAPTER 3

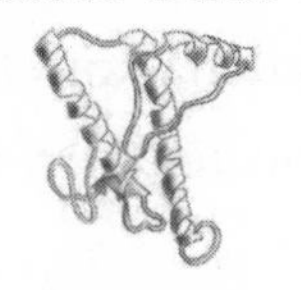

Protein

The Light of Dawn

I have smelt them, the death-bringers,
. . . and my bowels dissolve in the light of dawn.
T.S. Eliot, *Murder in the Cathedral*, Part II, scene 1

A PASTORAL INTRODUCTION

The genes contained within our DNA do their work through intermediaries, the proteins – literally the 'first things' – that are manufactured, at the bidding of our genes, by the molecular machinery within our cells. This chapter tells the story of a protein that turned traitor, through the fascinating though tortuous tale of its discovery, and the bitter experience of one of its victims.

Pete was an insurance clerk, unmarried, in his late twenties. Around the start of February 1998, when snow was lying on the hills, the north winds keen, and the days short, he began to feel afraid. He had no idea why. He had never experienced such a peculiar, gnawing sense of anxiety before. At times he felt that he was being hollowed out from within, as if something malign was gnawing at his innards; at others as if he were walking a blade, or teetering on a cliff edge.

He confided his fears to his diary, and to one or two friends – who told him not to be daft. He tried to follow their advice but the feelings didn't go.

More than two centuries before Pete began to worry, farmers around Europe had become alarmed by a fatal disorder among their sheep. According to a description, from 1759, the affected animals 'lie down, bite at their feet and legs, rub their backs against posts . . . and finally become lame. They drag themselves along, gradually become emaciated and die.' The disease was potentially ruinous for their owners. Sheep were central to the economy of Europe: the wool trade, in one way or another, involved 'nearly a quarter of the British population which at 10 million people approximated the number of British sheep'.

The paralysing ailment in the flocks gave rise to discussion in the British House of Commons in 1775. It was apparent that the disorder might spread through the flock: 'a shepherd who notices that one of his animals is suffering . . . should dispose of it quickly and slaughter it away from the manorial lands for consumption by the servants of the nobleman'. The disorder went by a host of names, coined by the farmers who described and suffered its troublesome consequences – goggles, rickets, turnsick, shakings, rubbers, scratchie, Cuddie trot. But the name that stuck was scrapie, chosen because affected animals scraped themselves incessantly against posts and fences. The affliction preyed on European flocks throughout the nineteenth century.

Proof that the disease was indeed infectious came in 1936. Two French vets, J. Cuillé and P.L. Chelle, published the results of experiments showing that if material from the brains of animals with scrapie was injected into the brains of healthy sheep, they developed the condition, in their turn, one to two years later. The brains of affected animals, whether naturally affected or artificially infected, underwent a curious 'spongiform' change, developing widespread microscopic cavities. Soon afterwards an unintended human experiment showed that injection directly into the brain was not essential. When sheep vaccinated against another neurological disorder, 'louping ill', went down with scrapie, it turned out that the vaccine had been prepared using material from scrapie-infected flocks: the infection must have made its way from the site of injection in the muscles to the brain. In the first half of the twentieth century scrapie remained a cause for concern to farmers, and a puzzle for veterinary scientists – the infectious agent was extremely elusive. But it seemed unlikely to pose any threat to man, given that we must have been eating scrapie-infected meat, every now and then, with apparent impunity for over a century.

Unlikely things can come to pass. The risk posed by the 'scrapie agent' to human health has gradually come to light over the past fifty years – partly through a series of tragic accidents, and partly because of an original line of research that has linked infection and heredity to the fundamental role of proteins in our lives.

THE GHOST WIND

By June Pete was nearing the end of his tether. He still felt desperately anxious, and now his right foot had begun to tingle – an odd sensation, as if it were constantly on the point of recovering from a local anaesthetic. He had been to see his doctor, and confessed eventually to worries about cancer and AIDS. His doctor had reassured him that there was no reason to think he had either. When Pete returned a few days later with much the same complaints, he was referred on to a psychologist. His doctor thought he might be offended at this suggestion, but Pete did not care whom he saw – if only they could make him feel normal again.

In the 1950s, a world away from Europe, a strange catastrophe befell the Fore Indians – a Stone Age tribe living in the Eastern Highlands of Papua New Guinea. J.R. MacArthur, the Australian administrative officer in the district, described what he had seen as he approached some native dwellings: 'I observed a small girl sitting down beside a fire. She was shivering violently, and her head was jerking from side to side. I was told that she was the victim of sorcery, and would continue this shivering, unable to eat, until death claimed her'. The girl's condition was called 'kuru' by her tribe, meaning a 'trembling' as if with cold or fear, a trembling they associated with the 'zona', the 'ghost wind' which sometimes blew across the land. But this shivering continued when the wind died down, progressing over a year or two to unsteadiness, paralysis and death. The girl MacArthur chanced upon was one of many similarly affected, but although the condition was gradually spreading through the population, it was curiously selective, picking out the women and children of the tribe, leaving the men more or less untouched.

In 1956 a young, energetic and wilful doctor, with an interest in unfamiliar cultures and a taste for scientific quandaries, visited Papua New Guinea from Australia, drawn by rumours of the unusual brain disease which was said to be spreading through the island. 'Paediatrician by training, virologist by experience, genius by nature', Carleton Gajdusek's genius was transfixed by the spectacle of kuru. Ignoring instructions to return to base – the Walter and Eliza Hall Institute for Medical Research in Melbourne – Gajdusek embarked on a painstaking study of kuru, eventually winning the support of his supervisor, the immunologist Sir Frank Macfarlane Burnet. By the time Gajdusek started work, kuru had become responsible for half of the deaths occurring among the Fore Indians. Gajdusek expected to find that the illness was an encephalitis – an inflammation of the substance of the brain, usually caused by infection – but at post-mortem he found no signs of inflammation whatsoever. Puzzled, he embarked on a 'kuru safari', travelling long distances on foot, documenting the history and distribution of the illness. He established that the disease was limited to the Fore territory and that it had emerged within the living memory of the tribe. He considered toxic, dietary and genetic causes, eventually coming to suspect the latter, though an inherited trait could not easily explain kuru's predilection for women and children, or the occasional case in a girl marrying into the tribe. He noted that, under the microscope, brains infected by kuru showed a peculiar spongiform change. There the mysterious matter of kuru rested for a while.

Three years later, in 1959, after encountering Gajdusek's description of the pathology of kuru, an American vet, W.J. Hadlow, pointed out similarities to the microscopic appearances of the brains of sheep suffering from scrapie. By this time scrapie was known to be infectious and transmissible in animals. Gajdusek therefore returned to New Guinea in 1962 to gather more material for the crucial experiment: would kuru prove infectious if injected into the brains of animals? Chimps inoculated with the material Gajdusek supplied from the brains of patients with kuru went down with a similar disease in 1965, after an incubation period of about two years. Although the chimps had received direct injections of material from kuru brains into their own, it turned out that, as with scrapie, this was not essential: the disorder could also be transmitted by feeding infected material to chimps.

Remarkably, it turned out that the consumption of infected brains was also the explanation for kuru. The onset of the disease among the Fore people had followed their adoption of a very particular practice: the cannibalism of recently deceased relatives as part of the funeral ritual. The brain, not considered a delicacy, was left for the women and children, while the men ate the muscle – hence the much higher rate of infection among the females of the tribe. No new case of kuru has been contracted since the Fore people abandoned cannibalism.

Gajdusek was awarded the Nobel Prize for Medicine in 1976 for his discovery. Although kuru had been the focus of his interest, he was convinced that the condition was not merely 'a minor problem of stone-age people. It is more than that and if solved will certainly give to medicine important new leads.' His transmission experiments rapidly proved him right.

CJD

Julia, the psychologist, was puzzled by Pete's case. Health anxieties are common enough – she saw a stream of people preoccupied by worries about illnesses they had been assured they did not have, but Pete was out of the usual run. He was not a worrier by nature, and there seemed to be no particular stressors at work in his life just now likely to turn him into one. He also mentioned problems with his memory which sounded worrying to Julia. She had spoken to the GP who referred him, promised to do her best, but suggested that they should keep an open mind about his problem.

Pete, meanwhile, was losing ground. In October, about eight months after his anxiety set in, he went home from work one day just minutes after arriving. He had forgotten the password for his computer and had been too embarrassed to ask anyone for help.

There was another reason for his early departure that morning – he had begun to suspect that his colleagues meant him harm. When the girl who worked opposite Pete looked up and smiled as he walked by, he felt she was hiding something – something that involved him. He wrote in his diary: 'My mind's not right. I don't know why. I can't remember anything.'

Besides injecting material from kuru-infected brains into chimps, Gajdusek injected material from people who had died from a variety of 'control' conditions. Most of these injections had no effect, but one other disorder, also associated with spongy change in the brain, turned out, like kuru, to be transmissible. Creutzfeldt-Jakob disease (CJD) had been described by two German physicians in the 1920s. Subsequent research has shown that CJD occurs consistently in about one in a million people per year wherever it has been studied. Although the changes it causes in the brain somewhat resemble those seen in kuru, the effects on the sufferer are rather different. Few other disorders dismantle the capacities for movement, memory and thought so rapidly and efficiently. The disease generally strikes in late middle age, progressing so fast, over the course of a few weeks, that by the time the diagnosis is suspected and the patient is seen by a neurologist, communication is often impossible. At this stage of the illness a tell-tale sign is usually present: a rhythmic jerking of the body, known as 'myoclonus', commanded by repetitive discharges from deep in the damaged brain.

The discovery that CJD could be transmitted, and was therefore infectious in some sense, raised more questions than it answered. Determined efforts to track down the source of infection in ordinary cases were unsuccessful. There were some promising leads: a cluster of cases among a group of Libyan Jews fond of consuming lightly cooked sheep's brain and eyes raised the possibility that CJD might be caused by infection with scrapie. This would have made reasonable sense. But these studies ultimately drew a blank: there was no convincing evidence that CJD was generally caused by infection. But rare cases, tragically and notoriously, clearly *did* follow infection – as a result of inadvertent medical use of infected material. Transplanted corneas, brain lining retrieved from cadavers and used for neurosurgical repairs, contaminated instruments and, most poignantly, growth hormone recovered from the pituitary glands harvested from the morgue and then used to treat children with growth hormone deficiency, have all transmitted the condition in a series of medical misadventures around the world.

Another observation only helped to wrap the mystery in an enigma. Whilst infection could sometimes lead to CJD, but was apparently not usually responsible, some cases, about one in ten, were clearly *inherited*. Families were described in which CJD was handed down from parent to child, like eye colour or Huntington's disease. Two other inherited conditions also turned out to be inherited 'spongiform' disorders. The first, Gerstmann-Straussler-Scheinker disease (GSS) is in fact a small family of conditions, giving rise to a variable mix of intellectual decline, unsteadiness and symptoms akin to those of Parkinson's disease. The second, a 'peculiar fatal disorder of sleep', Fatal Familial Insomnia (FFI), is arguably the most extraordinary of all the human spongiform diseases.

FFI was described in 1986 in a large – and distinguished – Italian lineage. Its first symptom is usually a progressive sleep disturbance with worsening insomnia. What sleep remains is punctuated by the enactment of dreams: the first reported sufferer was a 53-year-old whose sleep, lasting for only an hour, was 'frequently disturbed by vivid dreams, during which he would rise from his bed, stand and give a military salute. When he was awakened by his relatives he would report dreaming of attending a coronation.' As the months passed he became persistently confused, with a relentlessly worsening disturbance of the 'autonomic' functions, such as body temperature and blood pressure, that the brain normally handles without any conscious effort. He died nine months after the onset of his symptoms. Examination of the brain revealed damage centred in areas of the thalamus, a region roughly the size of a large nut, near the centre of the brain, which plays a key role – as one might have guessed – in the control of sleep, wakefulness, memory and autonomic functions. FFI is undoubtedly inherited in the handful of families it ravages, but like GSS and CJD it can *also* be transmitted from human to animal brains.

Making sense of a disorder which usually comes out of the blue, like the majority of cases of CJD, but that can also be caught from contaminated instruments and medicines, and yet is sometimes inherited, called for some revolutionary thinking.

ABSENCE OF PROOF

Absence of proof is not proof of absence.

Anon.

When I first met Pete, chatted with him and examined him, I had an experience, rare but memorable, that most doctors will recognise – I was sure that Pete was ill, sure that his illness was unusual, and sure that just now I was unable to give it a name. He complained that the light was too bright, and peered at me through half-closed eyes. It was possible to talk with him – he gave sensible, short answers to most of the questions I asked. But he seemed preoccupied, or distracted: I had the sense that his mind was otherwise occupied, though neither of us could have said what was occupying it. On a basic test of mental functions, he scored slightly, but definitely, lower than expected. The standard neurological exam – eye movements, muscle tone, reflexes and the like – indicated undoubted, widespread, but currently mild dysfunction of the brain. I arranged for Pete to travel to my ward, for some diagnostic thought.

In 1986, the year in which FFI was first described in the medical literature, vets in the south of England began to report a curious affliction of cattle. Affected animals became apprehensive, sometimes to the point of frenzy, and unduly sensitive to touch and sound; their heads dropped low, as if too heavy for their necks, they trembled, staggered and fell. Thirty years on from Gajdusek's pioneering work with kuru, the prepared minds of British scientists were poised to work swiftly. Two cow brains supplied to the Central Veterinary Laboratory at Weybridge showed spongy changes which were unheard of in cattle, but very reminiscent of what vets were used to seeing in the brains of sheep with scrapie. As the number of cases mounted over the following months there could be no doubt that an epidemic of a novel disease of cattle was under way in Britain, a spongiform disease of the brain which was christened 'bovine spongiform encephalopathy', or BSE, by the professionals and 'mad cow disease' by the press.

In 1988 it was made compulsory for farmers to report the presence of the disease in their herds. By 1996, 161,000 confirmed cases had been reported in the United Kingdom. The search for the cause of this new disease made public facts about animal farming which surprised most of us a lot. It turned out that before the practice was banned, in 1988, the feed of British livestock was regularly enriched with material from 'rendered' carcasses: to put it bluntly, the cows we thought were grass-eating vegetarians were also eating each other. This had allowed the epidemic spread of BSE, as infected animals were served up to other cattle. How had the infection entered the food chain in the first place? The source may never be known for sure. The likeliest candidate is a change in the procedure by which sheep carcasses were processed into cattle fodder, allowing scrapie-infected meat to reach cows in sufficient amounts to cross the species barrier and to set up infection in cattle (the 'species barrier' is relative: work with a number of species has shown that infection is less likely to occur when a member of one species is infected by a member of another – but clearly infection can still occur). Alternatively, BSE may have arisen spontaneously in a cow or bull whose brain was later recycled into foodstuff.

BSE was a disaster for British farmers – and regrettably we exported our misfortune round the world, in contaminated feed. Cattle were not the only victims: domestic cats and several captive species in zoos – including the lion in Edinburgh Zoo – succumbed to the plague. Consuming material infected with BSE was clearly an effective route of transmission among animals. What was more, the agent responsible was well able to jump from species to species. Despite this worrying evidence, the best scientific advice, through the late 1980s and early 1990s, was that the risk of transmission to *man* was remote. This advice was broadcast loudly and simplistically by the British government of the day, anxious to protect the interests of the beef industry. The British Minister of Agriculture at the time, John Gummer, was famously televised feeding a hamburger to his young daughter. By contrast, Stanley Prusiner, the leading American expert on this family of disorders, took a studiously noncommittal stance in a lecture given to London's Royal College of

Physicians in 1993: 'Whether BSE poses any risk to humans is unknown.' Whatever the scientists and politicians might have said in the 1990s would almost certainly have been too late to save Pete from his fate.

In 1986 I worked for a few months as a 'house physician' – a role also known affectionately as 'house dog' or 'house slave' – for a taciturn but widely respected and much loved Professor of Neurology, Bryan Matthews. He had a long-standing interest in CJD. Matthews had, in fact, supplied Carleton Gajdusek with one of the specimens which allowed him to prove that CJD was transmissible to animals. Matthews decided to have a go at pinning down the cause of CJD, and supervised a series of young neurologists in a 'case-control' study of the illness, designed to establish whether there was anything special about its victims – any quirk of their diets, duties or diversions which might provide a clue. This work contributed to the evidence that CJD was not caught, as a rule, like an ordinary infection from any identifiable source. It also ensured that there was a nucleus of interest and expertise in the condition among British neurologists.

In 1990 the Department of Health set up a small Surveillance Unit to monitor the incidence of spongiform disease of the brain in man, in view of the unlikely possibility that BSE might spread to humans. The expectation was that the surveillance scheme would draw a blank. Its director was Dr Bob Will, a neurologist in Edinburgh who had worked on Bryan Matthews's case-control study. The work of the Surveillance Unit was aided by the fact that neurology in Britain has remained a small speciality whose members tend to stay in close touch: Bob could be reasonably confident that unusual cases of dementia would be referred to the unit. So it proved.

Between March 1995 and January 1996 ten patients were referred to the Surveillance Unit with a distinctive story. All were young, most in their teens and twenties. All but one had consulted a psychiatrist before coming to neurological attention. Their reasons for seeing a psychiatrist were variously depression, agitation, aggression, anxiety and paranoia. Some sufferers noticed tingling in their limbs. But vague bodily complaints are common enough in anyone who is anxious or depressed. Only after six months or so of psychological upset or aching, tingly limbs did the true nature of the disorder reveal itself in unmistakable problems with memory, unsteadiness, falls, involuntary movements and incontinence. Death followed on average fourteen months after the first symptoms.

On 20 March 1996 Will and his colleagues announced their findings, to the genuine surprise of most scientists in the field. The account given in the day's news, which I was startled to hear as I drove home from work one evening, is one of my 'flashbulb memories' along with Kennedy's death and 9/11. Like the patients' case histories, the pathological changes identified in their brains by Bob Will and James Ironside, his pathologist collaborator, were highly distinctive. There was prominent spongy change, but also striking accumulations of protein in small dense deposits, 'plaques', similar to those found in two other conditions – scrapie and kuru.

This illness was something quite new: its victims were much younger than most sufferers from CJD; their illness ran a different, longer, course, with psychiatric symptoms at the outset; the changes in the brain set it apart. This unwelcome event was precisely what the Surveillance Unit had been established to look out for. The authors of Will's paper concluded cautiously that 'the most plausible explanation of their findings' was that the disease resulted from exposure to the cause of BSE. The discovery was alarming and depressing. It brought home the lesson, a hard one, that the absence of any proof of a risk does not equate with proof of absence of a risk. It was also scientifically impressive: the Surveillance Unit team had managed to detect a new disease occurring in ten patients in a nation of fifty-five million. But what *was* the cause of scrapie, kuru, CJD, BSE and their newly recognised, wholly unwelcome, British cousin, now known as 'v' (for variant) CJD?

FIRST THINGS

Protein, from *proteios*, Greek, 'of the first quality'

On the ward, a couple of days after first meeting Pete, I repeated my assessment. Again I was struck by his strange air of absorption – absorption by nothing that Pete could explain. Again I found the mild but consistent signs of disturbance in all the major brain systems controlling movement, systems that are much more commonly picked out singly by disorders of the brain. By that time Pete's brain had been scanned, revealing a subtle, symmetrical brightening of the thalamus, the deep-seated nut-sized region affected selectively in Fatal Familial Insomnia, and, as Bob Will's team had discovered, affected early in variant CJD. There could be little doubt that this novel, dire disorder was the most likely explanation for Pete's difficulties. I met with his family. They had heard about BSE. They knew that it could affect people who had eaten infected meat. They were shocked and scared. They wanted to know more about what caused this illness. We wondered what best to say to Pete.

The overwhelming evidence from kuru, scrapie and their human cousins, that these diseases could be transmitted from man to man, man to animal, animal to animal and animal to man had to mean that some infectious agent was involved. Work with the scrapie agent, years before, had shown that the agent must be very small: filters which would remove bacteria left its infectivity unscathed. This initially suggested that the agent was a virus. Viruses are essentially packets of genes which code the instructions for making more of themselves. They exist on the boundary between living and unliving matter: relying on the machinery of the cells they infect to reproduce themselves, they enjoy a tenuous 'borrowed life'. 'Slow viruses' of some kind seemed, for a while, reasonable candidates for the cause of scrapie and its relatives, given the small size of the infectious particle and the long interval between infection and the onset of disease. HIV, the causative agent of AIDS, is the most familiar of slow viruses. But there was a serious problem with this idea: experiments had shown repeatedly that treatments, like irradiation, extremes of heat and formaldehyde which reliably destroy genetic material, including viruses, *failed* to destroy the scrapie agent.

But if the agent was neither a bacterium nor a virus, what could it possibly be? There seemed to be no plausible alternative. Indeed, the observation that infectivity survived treatments which destroy genetic material was itself heretical. The first law of modern biology was that the reproduction of any organism, including any germ, depends upon the replication of nucleic acids like DNA: no DNA, no organism; no organism, no germ – and no infection. Yet infection there plainly was in scrapie, kuru, CJD. Squaring this circle – like finding an explanation for the peculiar combination of infection, inheritance and spontaneous occurrence observed in these disorders

– required some highly original thinking. Einstein pointed out that 'a theory should be as simple as possible, but no simpler': so far the simple theories had failed to deliver. But the answer was already available, for a scientist with the courage and imagination to trust it, in the experimental evidence. The purification of the scrapie agent from infected brains had revealed just one constituent – protein. The scrapie agent *had to be* a protein.

Proteins, which we encountered in the last chapter, are large molecules, composed of smaller ones called amino acids. Our genes – all 30,000 of them – spell out the structure of proteins: within the crucial parts of our DNA which are translated into proteins, as we saw earlier, each 'triplet' of bases specifies a corresponding amino acid; the succession of triplets in the gene specifies the succession of amino acids in the protein, known as its 'primary structure'. The amino acid strings within the protein often organise themselves into helical or sheet-like shapes; these are the protein's 'secondary structure'. Finally, under the unpredictable play of a huge number of short-range attractions and repulsions, the molecule folds itself into its final shape, its 'tertiary structure'. This has a crucial influence on its function.

Proteins are fundamental to our biology. They exist in a multitude of sizes and shapes, and play myriad roles within the body. It may help to have a rough sense of their scale: a smallish protein molecule might contain a couple of thousand atoms in 150 amino acids, a large one four times those numbers. A middling length for a protein molecule would be around a hundred times the size of a hydrogen atom. Some proteins, like collagen, are 'structural' – structural proteins supply the scaffolding of our bones, the fabric of our ligaments, the delicate internal architecture of our cells; others, 'enzymes', regulate the host of chemical reactions on which the life of every cell depends; some are exported from cells to convey signals to neighbours, as hormones or neurotransmitters, while yet others are inserted into the walls of our cells as receptors to detect the presence of these signals. Subtle modifications of the proteins spelled out by our 30,000 genes bring it about that our bodies contain about 100,000 distinct proteins. Like all the constituents of our bodies, proteins are constantly being 'turned over' – manufactured, harnessed, retrieved and destroyed.

But despite their prime importance in our biology, their diversity and dynamism, proteins cannot seed infections: they are just not the right kinds of things. They cannot divide or multiply: they are servants, not leaders; tools, not technicians – or so we thought.

A REQUIEM FOR PETE: HOW MANY?

When there are so many we shall have to mourn . . .
W.H. Auden, 'In Memory of Sigmund Freud'

Pete died ten months after making the last entry in his diary, at the age of 29, unable, by the end, to stand, sit, feed himself or speak – unable, almost certainly, to know what had become of him. This was, at least, a comfort to his family who had more or less lived at his bedside through the months of his descent into oblivion. At the end everyone was glad when death released him.

Pete had been an average guy: loved by his family, popular with his colleagues, fond of a drink and a smoke, with an eye for the women who often had an eye for him. At weekends he fished, in the streams near his home. He liked to feel the air around him, to let his mind fall still as the water in his favourite fishing pool. When the catch was sufficient, he prepared the fish himself and enjoyed cooking them for his family or his girlfriend. Otherwise he hadn't been too particular about his diet. But one day towards the end of the 1980s Pete's luck must have been out: a hamburger, or a meat pie, a sausage or a sauce had delivered an invisible but fatal dose of an unknown infection which eventually bedded down in his brain, and began its slow destructive work.

His case was typical of vCJD. A review of clinical features in the first hundred cases, published in 2002, largely confirmed the impressions gleaned from the first ten: psychiatric symptoms usually preceded neurological ones, and accordingly psychiatrists were most often the first specialists to be involved. Low mood, withdrawal from social contact, anxiety, irritability, insomnia and loss of interest in the sufferers' usual activities were the common early complaints. A minority noticed pain in the limbs, trunk or face. By six months, the underlying brain disorder had usually begun to make itself felt more obviously, with unsteadiness of gait and slurring of speech. Beyond six months the widespread involvement of brain systems that I detected in Pete becomes the rule, with death around thirteen months after the onset of symptoms.

Any young death is of course tragic. But something about Pete's illness was especially miserable: the baseless anxiety at the outset, building for a while to paranoia; the gradual erosion of memory and thought, too gradual to prevent Pete from discovering the truth about his illness; the lack of any suitable place for him to go once the diagnosis had been made; but worst of all the certain knowledge that this disease could have been avoided – that Pete had been poisoned. For this illness had the same outlandish source as kuru. It was the product of cannibalism, though in this case the cannibalism had taken place in our farms, where cow was fed on cow.

No one can know for certain how the scrapie agent reached the victims of 'human BSE'. The likeliest culprit is 'MRM', mechanically recovered meat, which was blasted out of carcasses in the abattoirs once all the conventional cuts had been taken. MRM typically includes a good deal of nervous tissue, which was highly infectious in animals harbouring BSE. It was used widely as a filler in cheap processed meat products, but found its way into items as unlikely as jars of baby food. The one straightforward moral of the BSE epidemic and its human counterpart is that it behoves us to pay close attention to the source of what we eat.

How many human deaths shall we have to mourn in the wake of BSE? In the early days of variant CJD it seemed possible that hundreds of thousands of cases might be dispatched by a plague as devastating as kuru at its peak among the Fore people. To date, ten years after the first cases, there have been fewer than 200 deaths, and the number of new cases has declined steeply. These could still be the vanguard of a larger epidemic, but it begins to look as though we – most of us – may have been lucky. But what, precisely, is it that we have escaped?

PRIONS

Neither flesh, nor fowl, nor good red herring.
Anon.

The 'scrapie agent', the substance which accumulates in the brains of victims of spongiform disorders, and which can transmit the infection to others, really does appear to be a protein. Stanley Prusiner, the American neurologist and biochemist who was awarded the

Nobel Prize for Medicine in 1997 for his work, christened this heretical, rule-breaking material, prion protein (the term 'prion' is a rough contraction of 'infectious protein'). His explanation for its capacity to transmit infection runs something like this.

Prion protein, in minute amounts, is a normal resident in the fatty membranes that enclose every cell in the brain. It is present at this moment in your brain and mine. Like every other protein, its manufacture is controlled by a gene, the 'prion gene'. Like every other protein, it is removed and replaced as it ages. Its function is uncertain. Molecules of prion protein happen to be slightly unstable, in this sense: very occasionally their three-dimensional shape can change in a way which renders them indigestible when the time comes for the cell to discard them. The change has been likened to 'turning a chiffon curtain into a Venetian blind'. The indigestible molecules have one additional, crucial, and fatal, property: when they encounter their digestible sisters they convert them into the indigestible form. This gives rise to a chain reaction, indigestible protein begetting indigestible protein. As new molecules are synthesised and transformed, they accumulate relentlessly, eventually wreaking havoc in the brain.

The rogue protein's ability to convert its healthy relations into the indigestible state explains how the condition can be transmitted from one brain to another: how prions come to be *infectious*, and can transmit disease. The explanation for the *inherited* forms of prion disorders is thought to be that those at risk produce an especially unstable version of the protein, predisposed to flip into the indigestible configuration. '*Spontaneous*' cases of CJD may result from the very rare, spontaneous, transformation of normal molecules to the indigestible – and infectious – form.

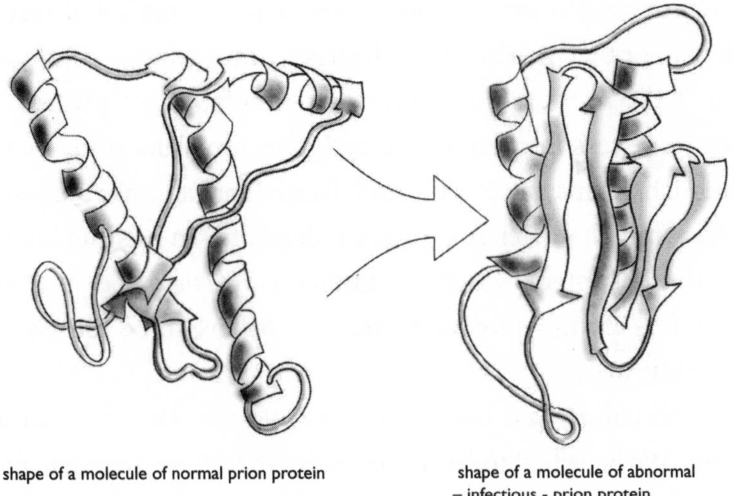

3. Prion Protein
The prion protein, a normal resident in our cells, is a string of amino acids – like all our proteins. In some places the string folds up into helical sections (the curly ones), and in other places into sheets (the arrowed ones). Normal prion protein can transform into an abnormal form which cannot be digested by processes that normally turn over the proteins in our cells: this form contains more sheets and fewer helices than the normal one. This abnormal form converts normal molecules into the abnormal form, leading to a chain reaction in which the indigestible protein progressively accumulates.

Prusiner's Nobel Prize award cited his discovery of 'a new biological principle of infection'. For anyone brought up with the familiar dogmas of modern biology, the idea of 'prions' is genuinely revolutionary, as much a leap of imagination as a product of observation (although, as an historical footnote, writing in the 1940s the physiologist Charles Sherrington described a gene as a 'self-fermenting protein'. Genes proved to be something different, but in the prion we have rediscovered precisely what Sherrington envisaged. Memories are short in science – as elsewhere).

The Nobel award is not, of course, a guarantee that the theory is watertight. Most scientists working on scrapie and its cousins have been convinced by Prusiner, though a few remain sceptical. It turns out that many other neurological disorders, much more common ones than CJD, result from a similar accumulation of protein – which for some reason cannot be recycled by the usual routes – in the brain: Alzheimer's and Parkinson's disease, for example, are both associated with characteristic protein deposits. In neither case, however, is the protein which accumulates *infectious.* So far, the prion disorders are the only ones which we believe to be caused by an infectious protein.

Success and failure have been mixed in equal parts in our dealing with prions. We have had to learn the lesson that these disorders are infectious and transmissible over and over again: for scrapie, for kuru and now for BSE and variant CJD. The best advice on BSE was wrong. Yet Bob Will's team was quick to spot the looming problem in the UK, and Stan Prusiner's imaginative science has made sense of two baffling puzzles: how a disorder can be sometimes inherited, sometimes infectious, sometimes occur spontaneously, and how a protein can transmit disease.

CHAPTER 4

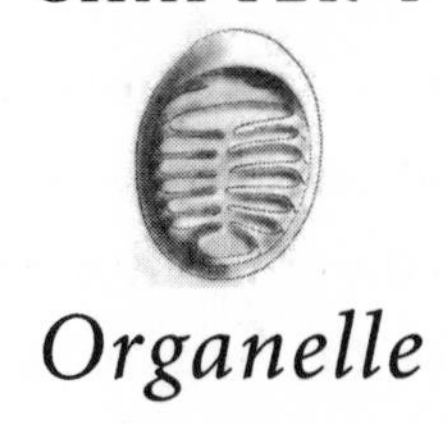

Organelle

Metamorphoses

All things move in order and all are in a state of flux.
Heraclitus

. . . bark began to thicken the smooth skin.
It gripped them and crept up above their knees.
They struggled like a storm in storm-tossed trees.
Then as each finger twigged and toe dug in,
Arms turned to oak boughs, thighs to oak, oak leaves
Matted their breasts and camouflaged their moves . . .
Seamus Heaney, 'Death of Orpheus',
from *After Ovid: New Metamorphoses*

ALL CHANGE

The proteins we encountered in the last chapter are mainly found within the cells of which our body is mostly made – though a few, like the hormones that course in our blood or the collagen bracing our bones, lie outside them. The following chapter will introduce the cell entire. But before we step back to admire it, we need to examine its chunkiest components, its 'organelles', biochemical robots built largely of proteins that dwell within our cells.

Their history, as it turns out, is one of extraordinary change. Change is everywhere and ceaseless, as Heraclitus pointed out, but some changes seem to require the five weighty syllables of 'metamorphosis' to do justice to their scale. Poets and artists have always been fascinated by metamorphoses, so I shall approach this shape-shifting topic obliquely, by taking you on a short trip to the museum.

Edinburgh's twin Galleries of Modern Art are housed, one opposite the other, in the handsome shells of two monumental eighteenth-century orphanages. They can be reached by a path beside a river, the Water of Leith, which threads its way from the Pentland Hills, on the city's edge, to Edinburgh's ancient port at Leith – once the haunt of rowdy sailors, now the focus of schemes of urban renewal. The Galleries face each other across green, undulating lawns, studded with the work of the sculptors Henry Moore and Eduardo Paolozzi. From their grounds one can gaze into the centre of the town, to the cathedral, the craggy castle on its hill and the distinctive, leonine profile of Arthur's Seat, Edinburgh's miniature urban mountain, rising above Holyrood Palace and the new parliament. On the scale of a human lifetime most of this scene is unchanging.

One of these sister galleries, the Dean, is the home of Edinburgh's collection of surrealist art, with paintings by Salvador Dali and René Magritte. Greeting the visitor, Paolozzi's magnificent Vulcan, a huge knobbly sculpture of the lame blacksmith god, strides across a room which only just contains him, a metallic giant soaring twenty feet to a metallic ceiling. But I keep being drawn back to a less famous work, which hangs across the corridor from Vulcan, Paul Delvaux's *The Call of the Night.*

Although it depicts a withered desert of mutilated trees and a few colourless, bare mountains, the first impression is of fertile abundance: in the foreground stands a naked woman, oval-eyed, full bodied, emphatically revealed to the viewer. She looks across the desolate scene with an ambiguous gesture, of acceptance or surrender, her head half turned over her left shoulder, her gaze averted from the viewer's, arms drawn back. But look again: this woman is becoming something other, changing before your eyes – her hair is tumbling foliage that has already taken root. Look beyond, to a second naked figure, gone further in her metamorphosis – or perhaps this time a tree is spawning woman. Another tree bears candles where there should be branches, leaves. At its root lies a sand-scoured skull.

Why linger at this strange scene? The lure of surrealism – like the literary genre of magical realism – is less mysterious than its subject matter. For we live in the midst of magic. Only a fraction of our experience is literal or mundane. In our dreams, the dead arise and greet us, strangers embrace us, trees walk, animals speak, familiar identities transform themselves before our unquestioning eyes. Even in the clear light of day, our experience is fraught with metaphor. Children, free of embarrassment, express this more openly than their elders: 'that tree's talking and that one's nodding its head', my four-year-old son reported, looking out at the wind-tossed leaves. But most of us, from time to time, sense the pressure exerted by our past on our perception of the present. 'Twice or thrice had I loved thee/before I knew thy face or name' wrote the poet John Donne in the seventeenth century – or, as Virginia Woolf's hero-heroine Orlando despairingly reports, 'everything, in fact, is something else'.

The shape-shifting quality of our dreams and the metaphorical character of experience help to explain the fascination of metamorphoses in literature and art. But there is another even deeper source. Change is the first inexorable law of life, and metamorphosis is a familiar, miraculous, feature of our biology. The growth of the single fertilised cell into the infant born nine months later, though very gradual, is just as extraordinary as any of the mythical transmutations Ovid describes in his wonderful *Metamorphoses.*

This chapter recounts two parallel tales of transformation. The first is of a disorder, or family of disorders, whose manifestations are so various, so peculiarly metamorphic, that its relatively recent recognition is understandable; the second concerns the microscopic cause of the disorder, itself the subject of an extraordinary metamorphosis, to which we owe nothing less than our lives.

LORNA

Lorna was one of seven, very much one of the gang, a cheerful and mischievous child. But even as a youngster an obvious difference marked her out from her three bonny sisters and three vigorous brothers: they were big, but she was little. All her brothers grew to be over six feet. As an adult, the shortest of her sisters was five foot seven; Lorna was only four eleven. She was as slight as she was small, nimble and full of laughter, with pale blue eyes and blond hair cut short in a fringe. She won the 100 metres at her town's Gala Day for two years running at the ages of nine and ten. The memory of this childhood triumph still brings a smile to her lips.

Like many children born into large, rumbustious families, Lorna took it for granted that she in her turn would be a mother. She married in her mid-twenties. Her husband, an affectionate, quiet, man, dark to her blond and not very much taller than Lorna herself, also wanted children. But fate seemed to be pitted against them. Eight years later Lorna had miscarried three times and delivered two babies so prematurely that they were too small and frail to survive more than a few days. Each time she conquered her grief and got back to work, quietly storing the sadness away somewhere secret.

Something else, far less poignant than the loss of her babies, had happened to Lorna since she married. She and her husband had completely failed to notice it, as we so often fail to notice gradual change, but it was spotted instantly by her sister returning from the States for the first time in years. Lorna's eyelids had begun to droop, just overlapping her pupils, giving her a permanently sleepy look, which was not disfiguring but a little disconcerting on a first encounter.

These droopy lids or 'ptosis' were also noticed promptly by the observant female obstetrician to whom she was referred for advice on her miscarriages. The specialist found that Lorna was mildly diabetic, her blood sugar rising excessively when she was given a sugary drink to test her glucose tolerance. Diabetes can interfere with pregnancy – but not, as a rule, so disastrously when the condition is so mild. This experienced doctor was puzzled: 'I wonder if she has some syndrome we are not recognising,' she wrote. As it turned out, she was absolutely right.

THE BIRTH OF CELLS

Two billion years ago, when the earth was roughly half its present age, it had been colonised widely by life, but by life of a distinctly alien kind. Like all of living creation, these primordial beings depended on nucleic acids to transmit genetic information, and on proteins to orchestrate chemical reactions and build bodies – but their bodies were simple in the extreme, microscopic packets of chemistry, the ancestors of contemporary bacteria. They are known to biologists as 'prokaryotes' because of the lack of internal compartments within their cells. Around 2.3–1.8 billion years ago a series of crucial developments created the kinds of cell from which all multicellular organisms are built, notably including you and me. These cells, termed 'eukaryotic', are complex and internally partitioned, and contain a range of distinctive inhabitants, known as the cell's organelles.

The details of the history that turned prokaryotes into eukaryotes are murky, but there is virtually unanimous agreement about one particular development, surely one of the most dramatic and far-reaching metamorphoses in biology. The ancestor of the cells from which our bodies are composed engulfed a bacterium. This bacterium proved able not only to thrive and reproduce itself within the cell that had consumed it, but also to help the cell itself to thrive. The free-living descendants of this bacterium survive to the present as the Rickettsia, the cause of the various forms of typhus. But within the engulfing cell, the bacterium was gradually transformed – over billions of generations – into an indispensable ally, locked by an ever more intricate chemical interdependence into the most intimate of relationships with its host. It became the cell's power station, the source of most of the energy the cell produces and consumes when oxygen abounds. In return the bacterium found a congenial home, where it had the opportunity to reproduce itself in tandem with its host, so that the evolutionary paths of host and parasite now joined a single track.

The descendants of this transformed bacterium are called mitochondria. They are found in every living plant and animal, with several hundred copies in each human cell. They are the most distinctive inhabitants of our eukaryotic cells. Take a moment to pause and consider. It is surely, to say the very least, remarkable that every one of the units from which you and I are built, every one of our cells, contains the descendants of bacteria that entered into symbiosis with our ancestors, and became their power houses, long before the dawning of multicellular life.

The highly adapted mitochondria found in bodies like ours are the final destination in the cell of the fuels that keep us going – principally sugars and fats. Discovering what happens next kept biochemists busy for decades in the mid-twentieth century. Within the mitochondria these molecules are efficiently stripped of their energy, which is harvested in the form of a molecule called ATP, adenosine triphosphate. The ATP produced serves as a kind of universal energy currency within the cell, available to fuel its many energy-consuming activities, like the building of its proteins. The highly regulated – but extremely rapid – burn within the mitochondria depends on the presence of oxygen, which combines with the carbon released from the burning fuels to form carbon dioxide, and with the hydrogen that is released as water, the 'water of metabolism'. The lethal properties of cyanide result from its ability to prevent the last in the intricate series of processes that hands hydrogen on to the waiting molecules of oxygen.

How can we be certain that mitochondria were once independent organisms? They still carry plentiful evidence of their origins. Most tellingly, they contain their own separate stock of genetic material, stored in a circular molecule that is recognisable as a bacterial 'genome', the organism's library of genes. They house the machinery that manufactures some of their own proteins – around 37 or so of the 3,000 different proteins present in a single

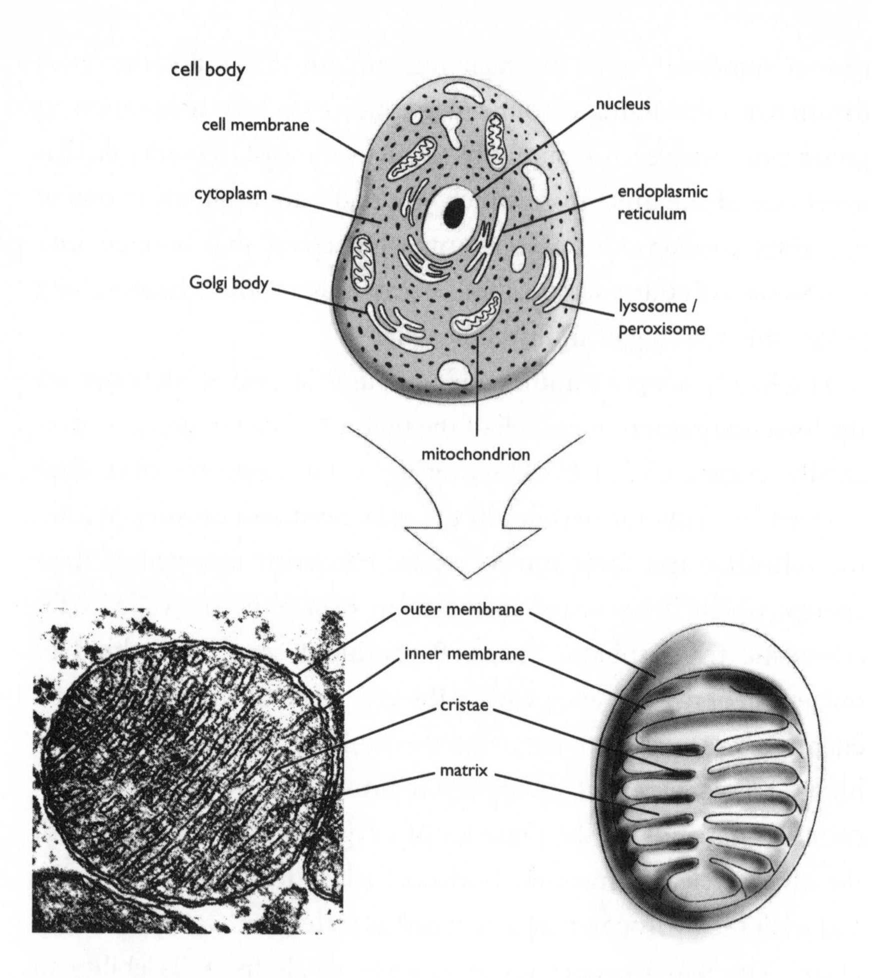

4. Organelles
The organelles found within the cell include the nucleus, which contains our DNA; the endoplasmic reticulum and Golgi bodies are involved in protein manufacture and processing; lysosomes and peroxisomes break down unwanted molecules; the mitochondrion, the hero of this chapter, is responsible for generating energy. The lower figure shows a mitochondrion magnified by an electron microscope to show its internal structure. The cytoplasm is the fluid interior of the cell.

mitochondrion – and they do so using a genetic code, and biochemical systems, subtly different from those used elsewhere in the cell. Finally, only mitochondria beget more of themselves: if a cell is separated from its complement of mitochondria it is unable to replenish them.

There is a curious medical footnote to this story. Aminoglycosides are powerful antibiotics used in the treatment of some severe infections. They target a structure found in all living cells, the ribosomes, microscopic production lines for proteins. Aminoglycosides are toxic to *bacterial* ribosomes – but because mitochondria have close affinities with bacteria, their ribosomes can be vulnerable too. The deafness that occasionally follows treatment with aminoglycosides in humans has been traced to the toxic effects of these antibiotics on our permanently resident, intracellular, bacterial associates – our mitochondria.

All the mitochondria found today in the cells of animals and plants are thought to have originated from a single ancient episode of fusion. But it is likely that there have been other comparable events in evolution. The ancestral cell that engulfed the precursor of our mitochondria may well, itself, have been the product of a fusion of an ancient 'archaebacterium' with a more progressive 'eubacterium'. Another coalition has been crucial to the lives of plants. Plant cells contain a second intracellular resident, with close resemblances to the mitochondrion: the chloroplast is the site of photosynthesis, the process by which energy from sunlight is harnessed to enable the plant cell to build sugars and starches from the carbon dioxide it captures from the air and the water it draws up from its roots. The evidence for these ancient coalitions have led Lynn

Margulis, the modern originator of symbiotic theory, to argue that Darwinian competition is only one of the principles of evolution: networking and symbiosis are also forces to reckon with. Cooperation sometimes counts for more than combat.

DUMBSTRUCK

Mercifully, Lorna defied the gloomy predictions of her doctors. After one more depressing miscarriage, she gave birth to a perfectly healthy and only slightly premature daughter, Iona. Lorna immensely enjoyed being a mother. But within a few years her own health had again become a concern. Her drooping eyelids began to get in the way of her vision. Her diabetes needed treatment with insulin. And she could not conceal from herself or from others that slowly but steadily she was going deaf.

Matters came to a head when Iona was eight. One day Lorna stood up in the kitchen after sweeping the floor, inadvertently banging the back of her head hard on the open door of a cupboard. The following day she noticed an intermittent flashing in her vision to the left. Over the next two days the flashing became continuous and was then replaced by darkness: she could not see in one half of space. She was told that the cause was a stroke – an interruption of the blood supply to a region of her brain. Her sight recovered over days. But as her vision returned, another, even more alarming, problem took its place. Lorna could no longer understand what others said to her, and when she spoke to them her choice of words was so bizarre that what she said was unintelligible.

When Lorna finally saw a neurologist, my predecessor, he recognised her underlying problem instantaneously. Some diagnoses are made painstakingly, by piecing together scattered scraps of evidence.

Others, like Lorna's, need no reflection if one has seen a case before. A favourite undergraduate teacher of mine called such diagnoses 'Aunt Minnies'. Recognising your Aunt Minnie when you meet her in the street doesn't, as a rule, require a meticulous process of deduction. Once seen, a diagnostic Aunt Minnie is never forgotten. Confronted with Lorna, small, slim and deaf, with diabetes and drooping eyelids, my colleague had immediately suspected that the source of Lorna's varied symptoms lay in a failure of her intracellular energy supply – in a dysfunction of her mitochondria.

The effects of mitochondrial failure are most marked in the tissues of the body that work hardest – at rest, the brain, and, on exertion, our muscles. Accordingly my colleague directed his attention to Lorna's muscles: the movements of her eyes were limited, and all of her muscles were mildly weak, reinforcing his suspicions. Microscopic examination of a snip of muscle taken under local anaesthetic from Lorna's thigh – a biopsy – confirmed them. Muscles are built from large numbers of elongated cells lying in parallel, their 'fibres'. Tucked in beneath the surface of Lorna's muscle fibres were accumulations of abnormal mitochondria which had multiplied there in their unsuccessful efforts to provide the cell with energy. These accumulations showed up under the standard light microscope, with the appropriate stain, as 'ragged red fibres'. Magnified 16,000 times, by the electron microscope, the mitochondria themselves were clearly abnormal, packed with improbably regular crystals.

Lorna's 'strokes', causing her loss of vision and then disordered language, were symptomatic of the energy starvation in her cells. Happily these mitochondrial strokes tend to behave more benignly than the standard kind: the brain tissue involved is not dead, as it would be if the artery supplying it had blocked, but simply exhausted because of failure of its energy supply. It can recover spontaneously over days or weeks. Lorna's language had returned to normal when I met her for the first time in the clinic.

THE CHILDREN OF EVE

Over the aeons since the ancestors of mitochondria first took up residence within their hosts, the activities of their descendants have become ever more intricately interwoven with the larger life of the cell. Many of the genes which direct the functioning of the mitochondrion now reside within the cell's nucleus, home to the vast majority of our genes. Yet the DNA within the human mitochondrion continues to play a crucial role: it codes for – contains the information needed to build – a number of the proteins that enable the mitochondrion to generate the energy it exports to the cell. It also spells out the instructions for building parts of the biochemical machinery that manufactures these proteins. Genetic testing of the DNA from Lorna's mitochondria revealed a change in a single base, altering a single letter of the genetic code, in one of the genes that, via an intermediary, guides amino acids into place in the proteins that the mitochondrion manufactures.

The change is now recognised as the genetic hallmark of a disorder described in 1984, christened the 'MELAS' syndrome. The letters of this rather forbidding acronym stand for: **m**yopathy, or weakness of muscle, as seen in Lorna's drooping eyelids; **e**ncephalopathy, widespread dysfunction of the brain causing confusion or intellectual decline, which Lorna has so far escaped; **l**actic **a**cidosis, an acidification of the blood that can occur at times when the supply of energy from mitochondria fails and glucose is broken down via the inefficient, 'anaerobic', route we use when we sprint; **s**troke-like episodes, of the kind that caused Lorna's visual loss and speech disturbance.

These problems are usually accompanied, as they are in Lorna's case, by a small frame, diabetes and deafness.

While Lorna's difficulties were clearly the result of a primary dysfunction of her mitochondria, the expression of a single mutation in their complement of genes, secondary dysfunction within our mitochondria may be an important element of several more common disorders. Parkinson's disease, for example, has been linked to mitochondrial dysfunction. An outbreak of severe, early-onset Parkinson's disease in the US some years ago among drug users turned out to be the result of the toxicity of the designer drug in question, MPTP, to mitochondria in brain cells.

Mitochondrial DNA, the DNA containing Lorna's minute mutation, has a couple of idiosyncrasies. Almost all of your genes exist in two copies, one donated by your mother, one by your father. If you are male, the Y chromosome is an exception: this comes from your father and is what makes you both male. But whether you are man or woman, boy or girl, all your *mitochondrial* genes came to you from your mother. Sperm, of course, are packed with mitochondria: they generate the energy that enables the sperm's tail to propel it powerfully towards the womb and its single, slim, chance of immortality. But at the moment of conception, only the sperm's DNA is injected into the maternal ovum, while all its mitochondria are left behind, outside the ovum, surplus to requirements. The embryo's mitochondria are therefore all supplied by the ovum – and by mum.

Its second idiosyncrasy is that mitochondrial DNA acquires mutations at a much higher rate – roughly ten times higher – than 'nuclear DNA', the DNA stored in the chromosomes of the cell's nucleus. This is thought to be due to a combination of the chemically reactive conditions within the mitochondrion, where the process of energy generation gives rise to potentially toxic by-products, and the mitochondrion's relatively poorly developed mechanisms for DNA protection and repair.

Between them, these idiosyncrasies have made mitochondrial DNA an important tool for biologists exploring relationships between the races of man. As all this DNA is maternally inherited, lines of descent and genetic relationship are relatively easy to trace by studying the mitochondrial genome. Nuclear DNA, by contrast, is constantly being rearranged as DNA 'crosses over' between maternal and paternal chromosomes when sperm and ova are produced. And as mitochondrial DNA accumulates mutations relatively rapidly, ethnic groups or species acquire a high frequency of characteristic genetic markers within their DNA. Research relying on these two factors has shown, for example, that all existing human groups can be traced back to an ancestor living in Africa 150,000–190,000 years ago, 'mitochondrial Eve', and that her descendants, our ancestors, migrated from Africa into Asia around 80,000 years ago.

Most of the mutations in mitochondrial DNA which help anthropologists to trace these lines of human descent are 'neutral polymorphisms' – minute genetic changes which alter the DNA sequences without having any appreciable effect on function. These polymorphic sites are found scattered throughout our DNA, which is, like everything else that lives, undergoing constant, if gradual, change. Not all mutations, of course, are neutral: the mutation within Lorna's mitochondrial DNA is eloquent, a curse on the children of Eve.

ORGANELLES

A place for everything and everything in its place . . .
Samuel Smiles, *Thrift*, 1875

Mitochondria are prominent among the organelles, the membrane-bound compartments found within the cell. These compartments distinguish our eukaryotic cells – found in all plants and animals – from the simpler prokaryotic cells of bacteria. In case you wonder, as I did, about the etymology of these curious words, 'eukaryotic' means roughly 'good nut', prokaryotic 'first nut'. The etymology of the word 'mitochondrion' itself is also classical, from *mitos*, thread, and *chondros*, cartilage: under the light microscope the mitochondrion, if seen at all, is thread-like; under the intense magnification of the electron microscope its complex inner structure – the mitochondrion is itself divided into a series of compartments – looks cartilaginous (or alternatively, in Steven Rose's more endearing simile, like 'a pile of custard-cream biscuits').

The mitochondrion is one member of the family of organelles that collaborate in the ceaseless work of the cell. The queen of all the organelles, the matriarch of the family, is the nucleus. It contains the DNA that directs the business of life, known as 'chromatin' because of the strong purple colouring it acquires when the cell is stained for microscopy. During normal cellular business, parts of the chromatin, those parts that contain the genes required for expression at any given time, are unfolded. They are the focus of the process by which DNA directs the synthesis of proteins: DNA is transcribed into messenger RNA, which then carries the genetic instructions for manufacturing proteins from the nucleus into the cell. The membrane enclosing the nucleus is accordingly punctured with pores to permit the free passage of molecules to and fro. The theory mentioned earlier, that our cells are a hybrid or chimera of two primitive ancestors, proposes that the nucleus and part of its biochemical machinery descend from an 'archaebacterium'; the rest of the cell – excluding the mitochondria, which joined the gang later – from a 'eubacterium'.

Outside the nucleus, in the 'cytoplasm' of the cell, a bevy of other organelles awaits the arrival of the instructions from the nucleus. Ribosomes provide the microscopic production lines that translate the messenger RNA arriving from the nucleus into the proteins that build our bodies and orchestrate the chemistry of the cell. A system of membranous surfaces and channels – the endoplasmic reticulum and the Golgi apparatus – helps to organise the ribosomes and channel their products to destinations inside and outside the cell. Another group of organelles, the lysosomes and peroxisomes, acts as cellular household cleanser cum medicine cupboard, best kept under lock and key. These organelles contain a range of enzymes that are highly toxic if released indiscriminately, but invaluable if used sparingly for the digestion of simple and complex molecules reaching the cell – including alcohol – and for the recycling of its time-expired components.

In the background, minute even under the electron microscope, a network of filaments and tubules creates a 'cytoskeleton', a delicate, fluid, scaffolding that supports the structure of the cell and provides another set of molecular pathways within it. *Fluidity* is the key to the behaviour of all the cell's organelles. Nothing is rigid in this microscopic landscape: the cell's constituents are all in motion and in flux. The heroes of this chapter, mitochondria, are no exception. They go where they are needed: recent work has shown that during learning they make their way to the synapses, the points of contact between nerve cells, to provide the energy required for remodelling connections. Recent experimental work depleting the mitochondria from these regions reduced the sprouting of connections between nerve cells on which our memory and our mental lives depend.

We could not live our highly co-evolved, cooperative, symbiotic lives without our complement of mitochondria. Lorna has battled through against the odds and the protean effects of her disorder. At the last count she was well, cheerful and busy, regardless of her diabetes, deafness and droopy lids. Her resilience, despite the single, nearly fatal, flaw harboured in the ancient genome of her mitochondria, is inspiring, an example to us all. But do not be deceived. Imperfection is the rule of life, not the exception. We all accumulate mutations in our genes, and in our mitochondria, as the years go by: they are one of the major determinants of ageing. Battling through is the best we can hope for, as each of us undergoes his or her own inescapable metamorphosis.

CHAPTER 5

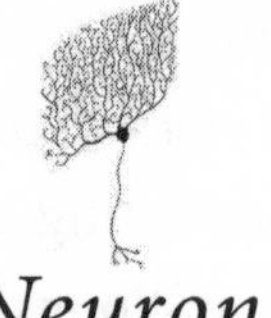

Neuron

Lost in Translation

LOST

Midway this way of life we're bound upon
I woke to find myself in a dark wood
Where the right road was wholly lost or gone . . .
Dante, *The Divine Comedy*

Ever been lost? Few of us can have escaped the desperate, deepening frustration of such moments: which turn leads from this maze? Which of these forking paths will rescue me from this dark wood? Worse, where is my child in the midst of this maddening crowd? Once in a while, we recreate these states of mind for fun – as kids, playing hide-and-seek, or as adults, cradling a detective story on our lap. We enjoy the tease of uncertainty, in the comforting knowledge that our perplexity is finite. Afterwards, home from the wood, the child retrieved, the mystery cracked, we forget the preceding struggle: though what seems so blindingly obvious now may have been far from obvious then. On occasions, we have to search and search – but often we need a good idea, a moment of insight, a feat of imagination. This chapter tells three parallel tales of things lost and found – cells in the brain pursued by scientists, cells lost in the brain of a patient, and a man's search for a way through the maze of his life.

At the start of the nineteenth century the nature and relationships of living tissues were deeply unclear. Robert Hooke, whom we encountered experimenting with air in Chapter 1, had famously observed a regular pattern in the slices of cork he examined with the first microscope as early as 1665, coining the term 'cell' as he did so, but the magnifications available to Hooke's followers over the next 150 years were inadequate to answer the questions that interested them most. Marie-François Bichat, one of the founders of histology, the study of living tissues, wrote in exasperation in 1801: 'The microscope is the kind of agent from which . . . anatomy never appears to me to derive much help'. But within forty years, improvements in the design of microscopes allowed some progress. In 1838 Gabriel Valentin glimpsed 'globules' with tail-like appendages in specimens taken from the brain. His supervisor, the prolific Czech-born biologist Jan Purkinje, speculated that the secret of life itself lay hidden in these 'granules', which he compared with the cells observed in plants. Others had similar hunches. A German lawyer turned medic, and finally botanist, Matthias Schleiden, had a lucky exchange of ideas with his friend, the microscopist Theodor Schwann. At the time, Schwann's work was focused on specimens taken from the spinal cord, the bundle of nerves within the spinal column that conveys signals concerning the limbs and body to and from the brain. As he gazed down Schwann's microscope, Schleiden noticed uncanny similarities between the plant cells he was studying and those visible in the spinal cord. In 1839, partly inspired by this insight, Schwann published his historic work: *Microscopical*

Researches into the Accordance in the Structure and Growth of Animals and Plants. In the treatise he proposed a 'cell-theory' by which cells were the building blocks of the tissues of both plants and animals – a 'universal principle of development' for living things of every kind.

He was right: cells are ubiquitous. They are the lowest common denominator of life. On close inspection with a microscope, they provide the underlying fabric of every organ of the human body – from our solid organs like skin, liver, kidney, muscle, heart and bone to our most conspicuous liquid organ, blood, where red cells, carrying oxygen, white cells, to fight off infection, and platelets, to aid the clotting of the blood, had been described by the 1850s. Cells vary greatly of course, from organ to organ, in their size and shape, and in the complement of genes that they switch on, sculpting their identity.

Schwann was right specifically that the cell provided a 'universal principle of development'. As the physiologist Charles Sherrington wrote many years later, 'to make a new beginning one must go back to the beginning'. During our first moments you and I were single cells, fertilised ova, giants of their kind one tenth of a millimetre in diameter, poised to embark on the quick-fire series of cell divisions that turns us into recognisably human embryos over the course of a few weeks.

Besides providing the building blocks for anatomy and development, cells are the fundamental units of metabolism and function. They can be coaxed, for example, into an independent existence in a laboratory's culture dish, where, if conditions are right, they continue to perform the role they played in the tissue from which they have been harvested. This is the basis for the recent flowering of stem cell research. Most dramatically, through the technology of cloning, individual cells can provide the seed for the creation of an identical twin organism, like Dolly, the famous sheep.

By the middle of the nineteenth century Schwann's cell theory was largely established for both plants and animals. But in one tissue of the body, the central nervous system – comprising the brain and spinal cord – the appearances glimpsed through the microscope continued to baffle their observers, giving rise to one of the major controversies of nineteenth-century science. The drawings made at the time, by the first microscopists of brain and spinal cord, depict a wealth of beautiful and improbable forms – tadpoles, starfish, surprised-looking one-eyed elephants – in the midst of a thicket of fibres, streaming, crisscrossing, interlaced. Some of these researchers assumed that the cell theory would eventually make sense of the exotic elements and their relationships. But others envisaged a unique structure for the brain, a system of elaborately interfusing parts, a 'widespread network of filaments anastomosing – interconnecting – one with the other'. There were good reasons in the background for both views: cell theory had opened up a fruitful approach to understanding 'the marvellous workmanship of life'; on the other hand, if the function of the nervous system was to allow communication and control throughout the body, then the concept of seamless interconnection was appealing. Puzzlement is writ large in the unwieldy vocabulary used by these early researchers, who spoke of the confusing, interlacing fibres as 'axis cylinder branches' and 'protoplasmic prolongations'. In truth, the details they most wanted to see, the course of the fibres, their branchings and, crucially, the points at which they met and perhaps fused, were simply invisible to them. There was no escape, for the time being, from the 'purgatory of unseen anastomoses'. They were lost.

THE SACRED DISEASE

This disease styled sacred comes from the same causes as others . . .
from cold, sun, and the changing restlessness of winds.
Hippocrates, *On the Sacred Disease*

The pretty town of Kirkby Lonsdale lies on a gentle slope above the River Lune. Upstream, the river flows through a quiet valley. In places its banks are precipitous and wooded, in others they roll more gently to the water's edge, dropping through green meadows to a pebbly beach. Raise your eyes higher, steeply, to the east, and they meet the craggier, pale outlines of the high dale, studded with sheep, transected by the occasional dark stone wall. To the west, across the Lune, beyond the town, the countryside undulates towards the grand, romantic outlines of the Lake District's peaks.

Long before I met him, Dave knew the contours of the valley in and out. As a child he had roamed the woods and played by the river, skimming the flat stones from the beach across its tranquil summer surface. Later, he fished its waters, or met up with his girlfriend in a secluded meadow close to his boyhood den in the trees. More recently, since leaving college, he had got to know the valley from a different perspective, driving a bus up and down its length half a dozen times each day, through the driving rains of winter and the beautiful, slanting sunsets of the summer.

Generally, these days, his attention was fully absorbed by keeping the bus on the road, snatches of conversation with his passengers and thoughts of the jobs awaiting him at home, where his riverside girlfriend was now in charge of a son aged two and a girl of five. He didn't generally have much time to admire the view. On this particular day, he was steering through a warm late summer's eve, the sunshine heightening all the colours of the dale to his left, creating a haze of golden light in the trees to his right. As always, he knew it was about to happen in advance.

Then it came. He felt the slight familiar churning in the stomach, a sensation that meant nothing and yet left him unnerved. Sometimes that was all. Dave told me that he was never quite sure whether he wanted it to end there, or half willed it to continue. Whichever, this evening the whole performance ran its course. As Dave steered his bus round the curves of the North Country lanes, the scene around him was transfigured: in an instant, he was weightless, an angel on the wing, soaring through a landscape fraught with significance, a sacred text that, for those moments and those only, Dave could read with fluent clarity. He was blessed. The fields, hills, river, sky and the whole valley sang in joyous celebration of the day. Who knows what Dave could or would have said if one of his passengers had tried to speak to him? Then, when the glorious significance of the scene was about to overwhelm him once and for all, it was gone, and he was Dave again. The experience lasted no more than a few seconds: its intensity was out of all proportion to its length. He knew it backwards. It used to excite and worry him, but now he just ignored it, and drove on, though as usual he felt somehow emptied out by what had happened.

He didn't bother to mention it at home. He had told his wife about the recurring experience once, as a curiosity, a wrinkle on the placid surface of experience. Alice had been interested, remembered some vaguely similar experiences of her own: long-forgotten moments in her teens, on holiday, when she had felt as if she were being somehow absorbed into a landscape she was wandering through contentedly, and other times when, short of sleep, she had felt oddly divorced from her surroundings. They had compared notes for half an hour and said no more.

But that night was different. Half an hour after they had dropped off to sleep, Dave's wife was woken by a moan from her husband's sleeping form, as if something had crushed him while he lay. When she turned towards him, she could just make out, in the half-light from the window, that his jaw was clenched tight, his body rigid. She had the first thought to occur to most people at these moments – that he was dying. But almost at once the stiffness passed, and a shaking replaced it, a monstrous series of convulsions that seemed to have nothing at all to do with Dave, as if some mischievous spirit had taken possession of his body and was playing, like a puppeteer, with its joints. After a minute that seemed very much longer, the shaking slowed and stopped. Dave lay pale and still, a trickle of blood running from his mouth on to his cheek. Alice pulled him on to his side and ran from the room for help.

DUMBFOUNDED

A look was enough. Dumbfounded I could not
take my eye from the microscope.
Ramón y Cajal

The discovery of the neuron, and with it the solution to the problem of the composition of the brain, emerged from the work of two extraordinary men, an Italian, Camillo Golgi, whose organelle we encountered in the last chapter, and a Spaniard, Don Santiago Ramón y Cajal. They were, in a sense, collaborators: Golgi, early in his long, immensely productive career, made a serendipitous discovery that later allowed Cajal, the younger by nine years, to tease out the anatomy and relationships of the elements of the brain with unprecedented clarity. Yet precisely the same observations led Golgi and Cajal to irreconcilably different conclusions. They met for the first time in Stockholm in December 1906, to receive the Nobel Prize for Medicine jointly. The occasion was not a happy one, as they made their fundamental disagreement plain in lectures of acceptance given on successive days. Cajal later described his indignation at Golgi's 'display of pride and self-worship'.

Camillo Golgi, the son of a doctor, was born in a small village in the Italian Alps. In 1843, he moved as a child to Pavia, where he qualified in medicine in 1865. His intellectual fire was kindled by Cesare Lombroso, a psychiatrist with an interest in the physical basis of mental disorders, and by teachers and friends with expertise in microscopy. In the end, this proved to have been a wonderfully productive combination of influences. But his early career did not go smoothly, and in 1872 financial necessity forced him to leave Pavia, to take a job as the resident physician in a 'Home for Incurables' in Abbiategrasso, a small town nearby. There, undeterred, he continued his work and made his discovery of the 'black reaction' or 'silver stain'. Like many good things, the discovery sprang from the most unpromising circumstances, as he worked at home, 'cut off from every form of scientific activity', in the kitchen, by candlelight.

In 1873 Golgi published his first report of this work, 'Sulla struttura della griglia del cervello' (on the structure of the grey matter of the brain). The real significance of the paper is its – admittedly brief – description of a new method for staining blocks or slices taken from the brain using a solution of silver nitrate. For reasons that remain uncertain to this day, the 'Golgi stain' is taken up by just a small proportion of brain cells, but it stains these in their entirety.

Under the microscope, the stained cells stand out against the dense surrounding background, like trees in bloom.

This crucial discovery did not immediately settle the dispute between those who believed that the nervous system, like the rest of the body, consisted of independent cells, and those who thought it was a continuous network. The finest ramifications of the nerve cells stained by Golgi's new method were still too small to be seen clearly using the light microscope – and, as the contemporary neuroscientist Gordon Shepherd remarks, 'unfortunately Golgi's imagination was most stimulated by those aspects he could see least well'. He jumped to two major, erroneous, conclusions: that the single 'axis cylinders' emerging from the cells he stained unite to form a fused network of fibres, and that the cells' bushy crowns of branches and twigs were primarily 'nutritive', like the roots of a tree. Golgi moved on to other and varied interests: he gave his name to the 'Golgi apparatus' mentioned in the previous chapter, to a sense organ signalling stretch in muscles, the 'Golgi tendon organ', and did important early work on the parasite causing malaria. He was rector of his university and a senator in Rome. Perhaps he can be forgiven for failing to find the time to revise his view of the relationships of nerve cells – but he was not forgiven by Cajal.

Ramón y Cajal, born in 1852, was also the son of a doctor, a village surgeon in Petilla, close to the Spanish border with France. As a boy, Ramon was mischievous, rebellious, restless and solitary. In regular trouble at school, resentful at the impositions of childhood, his greatest pleasures were roaming the countryside and sketching. He was temporarily apprenticed to a shoemaker and a barber, but in the end his father, who had academic ambitions of his own, inspired in Ramon a fascination with anatomy, leaving his son with a lasting love for the 'true masterpiece of Nature' – man. Like many energetic people, Cajal's enthusiasms were legion – he wrote poetry and a novel, became absorbed by gymnastics, developing 'monstrous pectorals', and then by metaphysics, reading widely in philosophy. He qualified in medicine in 1873, aged 21, in Zaragoza. Like Golgi, Cajal had a difficult early career. In Cajal's case it was also fraught with bouts of near-fatal illness – malaria, tuberculosis and depression – but he must have been extraordinarily resilient and determined. In 1883, at the age of 31, he was appointed to the Chair of Anatomy at Valencia. By then his consuming interest, kindled on a visit to Madrid to study for an exam, was microscopy. A later trip, in 1887, introduced him to Golgi's new method, an apocalyptic moment he described with memorable intensity:

> Against a clear background stood black threadlets, some slender and smooth, some thick and thorny, in a pattern punctuated by small dense spots, stellate or fusiform. All was sharp as a sketch with Chinese ink on transparent Japanese paper. And to think that that was the same tissue which when stained with carmine or logwood left the eye in a tangled thicket where sight may stare and grope for ever fruitlessly, baffled in its effort to unravel confusion and lost for ever in twilit doubt. Here, on the contrary, all was clear and plain as a diagram. Dumbfounded, I could not take my eye from the microscope.

THE SECRET ARBORETUM

You are a child of the universe, no less than the trees and the stars . . .
From an inscription in Old St Paul's Church, Baltimore, 1696

The cloud that curls round the eye of the storm mirrors the spiral membrane coiled within your ear, the spiralling shell of the snail, the unimaginable vastness of the spiral galaxies. The turbulent eddies in the shallows of a river echo the turbulence of flowing blood as our arteries fork, the swirling of rock from erupting volcanoes, the streaming jets of gas thrown out by dying stars into light years of emptiness. These resonances reach across the tracts of space and time. They remind us vividly that we are 'children of the universe, no less than the trees and the stars'.

Golgi, Cajal and their colleagues were struck by another set of visual echoes as they gazed through their microscopes at the denizens of the brain. It was impossible to look at the shapes outlined by Golgi's stain without being reminded of a garden or a wood. There was an abundance of trunks, branches, twigs, thorns, roots, ivy and moss: Cajal even found hyacinths. The details of the landscape varied: in places a 'tangled thicket' or 'dense forest' does indeed come to mind; in others the disciplined regularity of a town park, or, when a well-developed single cell stands proud, the handsome beauty of an oak outlined on the horizon. But where Golgi saw 'invisible anastamoses', Cajal saw independent 'elements', 'autonomous cantons' – *cells*! Cajal could no more visualise the points of contact between them than could Golgi, but he drew the correct inference from the highly selective properties of Golgi's stain: that it was filling discrete units which must therefore communicate not by 'continuity' but by 'contiguity', through some form of 'influence from a distance'.

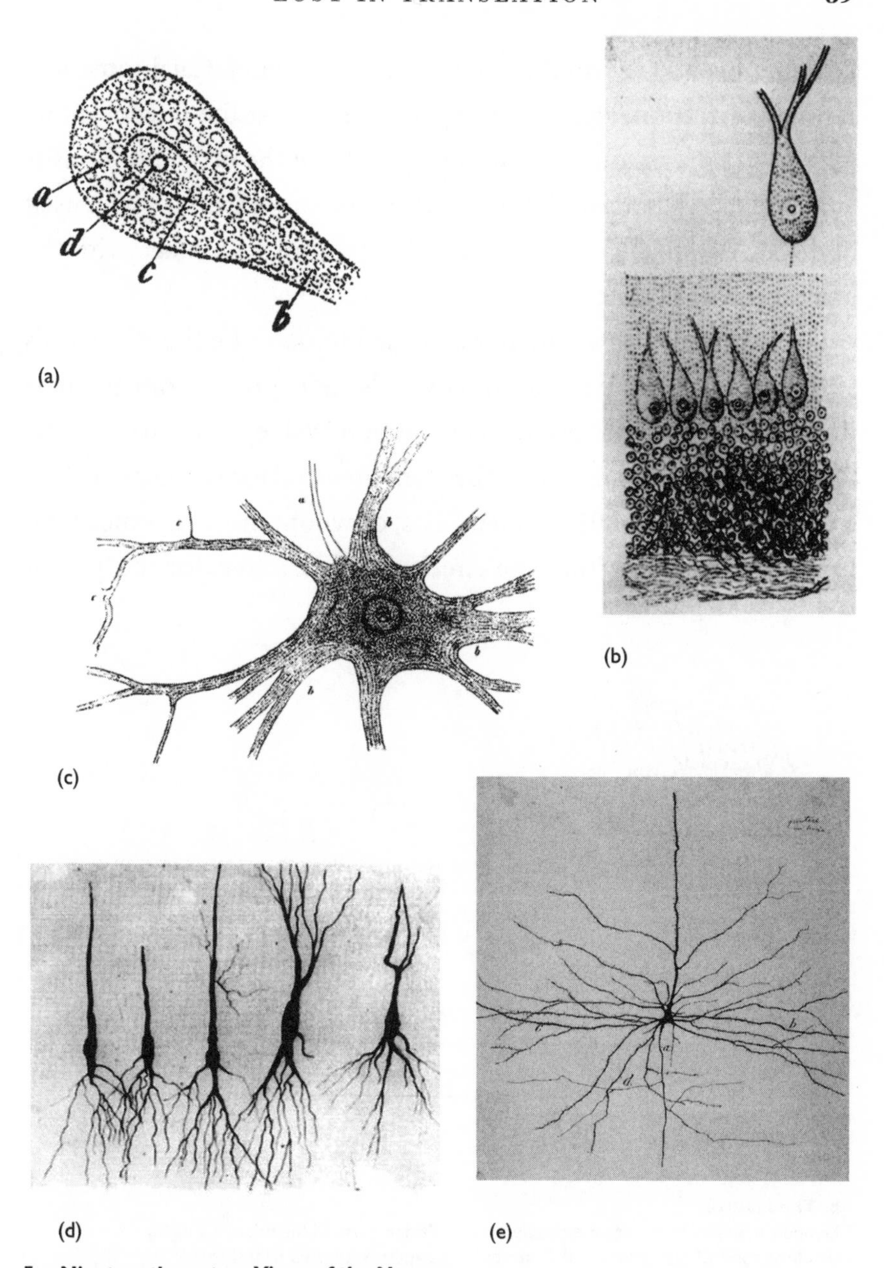

5a. Nineteenth-century Views of the Neuron

(a) the first clear microscopic image of a nerve cell, probably a Purkinje cell from the human cerebellum (Valentin1836).

(b) the first identified nerve cell in the nervous system: the large corpuscles of the cerebellum, which became known as Purkinje cells after their discoverer (Purkinje 1837).

(c) Kolliker's drawing of a large nerve cell from the anterior horn of the spinal cord, 1867

(d) Golgi image (reference to be confirmed).

(e) drawing of a neuron by Cajal, 1899.

Working his way round the brain, region by region, Cajal came to a second conclusion, again contrary to Golgi's: that the flow of information in the nervous system must pass from the nerve cell's bushy crown to its cell body and then on down the single trunk, which itself branches eventually, before making 'very intimate contact' with the bushy crowns of other cells.

Gradually, the unwieldy phrases used to describe the nerve cells and their parts were replaced by words that sprang from the new understanding of the brain. Wilhelm von Waldeyer, an aristocratic German biologist, suggested the term 'neuron' for the nerve cell in a review paper in 1891. Wilhelm His, of whom more in a moment, provided 'dendrite', from the Greek *dendron*, a tree, for the 'proto-

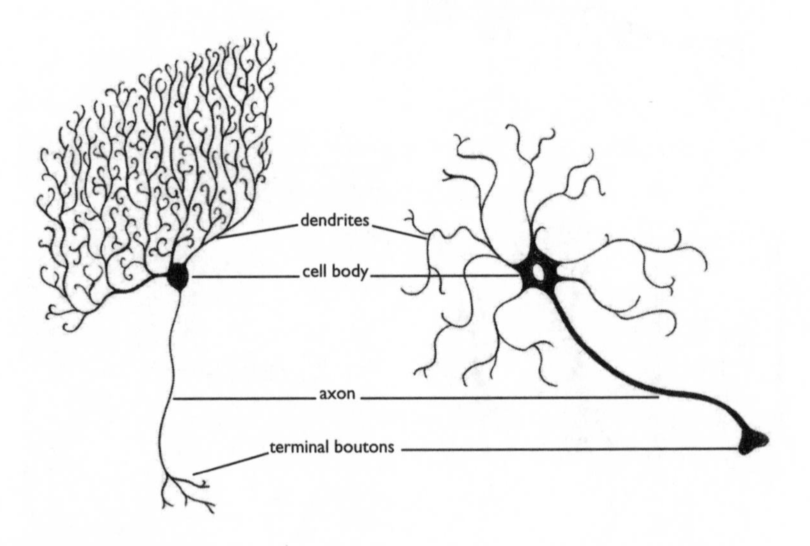

5b. The Neuron
The neuron, shown here in two examples from different parts of the brain, is a highly specialised form of the generic cell illustrated in Chapter 4, adapted to transmit information from one place to another. Its dendrites are its information-gathering antennae, picking up signals from other neurons. The cell body contains the same organelles and biochemical machinery as other cells. The axon – which can be two feet long - conveys the electrical signal away from the cell, for transmission to the dendrites of other neurons. The entire neuron is alive.

plasmic prolongations' of the neuron, the tufty crown of processes that receive incoming signals. The fibre carrying signals away from the cell was termed the 'axon', from the Greek for axle, by the grandfather of German cell biology, Albrecht Kölliker, by then in his eightieth year.

Names are much more than labels: we need them to articulate the theories we develop to make sense of the world, and when we use them we invoke these theories, summoning their powers. The theories in question are always imaginative, always go beyond the information given. Cajal's hypothesis – that the nervous system is created from potentially independent units, which gather information through their dendritic trees, transmit a signal down their axons and communicate across an intervening gap – was hugely clarificatory and productive. The work it inspired over more than a century has shown that it oversimplifies the truth – some neurons *are* in fact directly interconnected as Golgi had proposed, and information *sometimes* flows through the neuron in the direction opposite to the one Cajal envisaged – but Cajal's theory was challenging, testable, and made sense of a wealth of facts about the brain.

ECSTASY

A happiness . . . of which other people have no conception.
Dostoevsky

Dave didn't come round properly until he reached the hospital. It is a dismaying experience, to wake from your sleep several miles from the bed you climbed into, on a hard stretcher in a brightly lit cubicle, scrutinised by unfamiliar faces. Dave rolled on to a shoulder that turned out to be sore. As his eyes became accustomed to the light, he found Alice, standing in the doorway. She smiled. As he began to speak, he realised that his tongue was thick and painful. Alice came across and took his hand. He closed his eyes again and discovered, as he rolled back on to the trolley, that just about everything hurt.

Alice explained about Dave's seizure. The doctors were arranging a brain scan. Dave felt bewildered for a moment, then tried to piece together what had happened. For a couple of minutes his memory was absolutely blank. Then he remembered the feeling at work, and the subliminal, sickening thought that he might lose control of the bus. He explained, and Alice frowned. The casualty officer had told her that Dave wouldn't be driving a bus again for a long while. She passed on this news. Dave suddenly felt very bleak. He could cope with a sore tongue and an aching body, but how could he, or his family, cope without his job?

The nineteenth-century fathers of neurology, the contemporaries of Golgi and Cajal – perhaps even Hippocrates himself – would have considered a diagnosis of epilepsy, the 'sacred disease', on the basis of Dave's story. Epilepsy is the most common, serious neurological disorder. The risk of having an epileptic seizure over a lifetime approaches one in ten. At any time almost one in a hundred of us suffer from the recurring tendency to have such seizures that defines epilepsy. Such a common disorder must reflect important principles of brain function. Unlike the founding fathers of neurology, we know, nowadays, roughly what happens in the brain during a seizure.

In the ordinary way, in the waking brain, electrical signals travel constantly hither and thither, conveying and transforming the information that allows us to perceive, think and act. The corresponding wave forms in the EEG, which scans the electrical activity that penetrates to the surface of the scalp, are rapid and minute. This activity is often rhythmic in part, but in ordinary wakefulness these rhythms are harmonious elements of a complex whole. In epilepsy, this electrical activity synchronises markedly, disrupting the more subtle processes that underlie our mental lives. If this synchronisation involves the whole brain, consciousness is lost, and a 'grand mal' seizure can ensue, with stiffening of the body followed by convulsive shaking of the limbs – often accompanied by incontinence and tongue-biting. In the EEG the corresponding activity is conspicuous, pulsating and widespread. But if the epileptic disturbance disrupts the activity of only a part of the brain, it can manifest itself in much more subtle disturbances of experience or behaviour. Attacks like Dave's, of altered awareness, brief, intense and paroxysmal, charged with a numinous significance, are among the less common varieties of 'partial' epilepsy. They have been called 'ecstatic seizures'.

Fyodor Dostoevsky, the nineteenth-century author of novels *Crime and Punishment*, *The Brothers Karamazov* and *The Idiot*, suffered from epilepsy with auras of this kind. He described such attacks in the hero of *The Idiot*, Prince Myshkin:

> there was a moment or two in his epileptic condition almost before the fit itself (if it occurred in waking hours) when suddenly amid the sadness, spiritual darkness and depression, his brain seemed to catch fire. . . . His mind and heart were flooded by a dazzling light. All his agitation, doubts and worries, seemed composed in a twinkling, culminating in a great calm, full of understanding . . . but these moments, these glimmerings were still but a premonition of that final second (never more than a second) with which the seizure itself began. That second was, of course, unbearable.

Dostoevsky said, of his own seizures, that they produced 'a happiness that is impossible in an ordinary state, and of which other people have no conception. I feel full harmony in myself and in the whole world, and the feeling is so strong and sweet that for a few seconds of such bliss one could give up ten years of life, perhaps all of life.'

In the nineteenth century the underlying cause of such seizures would almost always have been unknown. A century later, Dave's scan explained his seizures, providing an explanation that would have fascinated Golgi and Cajal. It requires a few background facts about the growth of neurons.

THE LONG MARCH

One of the crucial clues indicating that the nervous system did indeed consist of independent cells came from reseach on its development. Studies of the developing nervous system of chicks persuaded the nineteenth-century German biologist Wilhelm His that neurons sited in the spinal cord sent probing outgrowths, axons, into the chick's developing limbs: these would later provide the nerves that activate the muscles of the wings and legs. He wrote: 'As a firm principle I . . . put forward the proposition: that each nerve fibre originates as an extension from a single cell. This is its genetic, nutritive and functional centre'. Cajal, seeing similar appearances some years later, could almost hallucinate their movement: the nerve ending 'could be compared to a living battering ram, soft and flexible'; it 'appeared [to be] endowed with amoeboid movements'. This work launched a new subject, neuroembryology, which aimed to describe and understand the development of the brain, the most intricately formed, elaborately interconnected, system that science has so far encountered. It is no surprise that, amidst the teeming confusion of the growing brain, neurons sometimes get lost.

The nerve cells that later form the cortex, the wrinkled outer mantle of the brain, a region closely implicated in the subtleties of our experience and behaviour, are born deep down, from stem cells hugging the walls of the ventricles, the fluid filled spaces that occupy the centre of the brain. The cells born there must travel, guided by radial support cells that extend like climbers' ropes, from ventricle to cortex, navigating amongst their companions, through a landscape of immense complexity that is everywhere changing its geography as growing axons probe, sniff and elbow their way towards their targets. It is as if you had to crawl for a couple of kilometres through a series of living tunnels, *en route* to a destination you have never visited before and have to find by scent and feel. Anything that can go wrong will do so, once in a while. Occasionally clusters of neurons fail to get started out on their journey at all: these 'cell rests' can be spotted on brain scans as thickenings of the ventricle's wall. Sometimes clusters of cells get waylaid *en route*; these are visible as clumps of grey matter lying in amongst the white matter – the fibres communicating between different brain regions – that they should have traversed on their long march to the cortex.

These misdirected neurons misbehave: rather than contributing to the orderly work of the brain, they transmit signals of electrical distress, pulsating cries for help. This is a potent cause of epilepsy; a cause that usually went undetected until advances in techniques for imaging the brain – in particular the development of magnetic resonance imaging, or MRI – made it possible to spot the small clusters of misplaced cells. Dave's brain scan was performed the day following his fit. During my conversation with him later in the day, he had to try to make sense of the unfamiliar facts of neuronal migration, because, as it turned out, a substantial detachment of his neurons had parted company from their comrades on the royal road to the cortex, and had been fomenting trouble in his brain.

BORDER CROSSINGS

The sea is the land's edge also . . .
T.S. Eliot, 'The Dry Salvages', *Four Quartets*

During the nineteenth century Golgi, Cajal, His and their colleagues made great progress in understanding how the brain was put together. But even Cajal, whose theory of the neuron still dominates neuroscience, had only the sketchiest grasp of what neurons actually do. He ended a lecture to the Royal Society of London in 1894 by acknowledging that the meaning of the 'dynamic ensemble' of nerve cells he had described was 'currently undecipherable'.

The brain was evidently responsible, somehow, for command and control, mediating and coordinating the workings of the body, experience and behaviour. It seemed likely that the cells with which it was packed conducted electrical signals; it had been known since the mid-nineteenth century that the nerves outside the brain did so. But it was not until well into the twentieth century that the details of the process began to become clear.

In brief, like all cells, neurons behave like minute batteries. They accumulate a tiny electrical charge, with a higher concentration of negatively charged atoms and molecules inside the cell than outside it. This creates an opportunity for signalling. When the difference in charge between the inside and the outside of a neuron is reduced by a certain small amount, molecular gates in the cell membrane suddenly open, allowing a rush of positively charged sodium atoms to enter the cell. Their entry reduces the negative charge in the adjacent segment of the neuron, the gates surrounding this segment open in turn, and a wave of sodium influx travels down the output cable transmitting signals away from the cell, its axon. Here is the basis for the neuronal 'action potential', the all or nothing, binary signal that conveys the neuron's crucial decision about whether or not to fire. Over the following few fractions of a second, a second set of gates opens, positively charged atoms of potassium flow out of the cell, and the normal electrical balance is restored.

This much-simplified explanation may raise a few questions in your mind. First, if what I have said is true, over time the levels of sodium in the cell must rise and levels of potassium must fall. So they would, were it not for a set of molecular pumps in the wall of the cell that prevent this from occurring by constantly exchanging sodium within the cell for potassium outside it. These require an energy supply – provided by ATP, from mitochondria – so that neuronal signalling eventually ceases if the cell's metabolism is blocked. Second, the description in the last paragraph assumed that *something* starts the process of signalling in the first place, but I did not explain what kind of something is involved. The answer is fundamental to the job that neurons do. The 'something' can vary from the arrival of light in the eye, sound in the ear or a taste on the tongue to the arrival of signals from other neurons, sometimes from many thousands of them, some increasing, others lowering the neuron's tendency to fire. Neurons are therefore 'transducers', capable of converting energy of one kind into energy of another, and 'integrators', summing influences from elsewhere. Their integrative function is one of the keys to their role in command and control. Since they don't, as a rule, make direct contact with one another – as

Cajal correctly insisted – the electrical signal travelling down the axon has to be conveyed by a different kind of messenger from one cell to the next. But that is another story, for which I will keep you waiting until the next chapter.

Axons are not just bionic cables, transmitting electrical signals round the brain and body. They are living extensions of the cell, conveying vital materials to the furthest reaches of the neuron: stem the flow, and the axon will bulge at the obstruction, just as the branch of a tree bulges at a ligature. But for Dave, and his 'sacred disease', the vital question was how to calm the uninhibited firing, the electrical cries of distress, from the squadron of lost cells within his brain.

FOUND

... the end of all our exploring
Will be to arrive where we started
And know the place for the first time.
T.S. Eliot, 'Little Gidding', *Four Quartets*

Drugs are more often discovered by chance than built by design. At least until very recently, we have known too little about the minute workings of the body, and have had too little skill in manipulating molecules, to construct drugs from first principles. But we do now have a grasp of many major mechanisms of drug action, understanding why they work even if we found them through good luck.

The most widely used of the dozen or so drugs on offer for Dave's kind of epilepsy works by reducing the firing of neurons, especially neurons that are exceptionally – or excessively – active. Taken in overdose these drugs depress the functioning of the whole brain. But it is often possible to find a happy medium, a dose that will calm the paroxysms of electrical discharge that cause seizures without preventing the electrical activity on which our mental lives depend. So it proved, after a few months spent finding the optimal dose, in Dave's case. He has never had another major seizure. Occasionally, but often enough to prevent him from getting back behind the wheel, he still experiences his aura, the uneasiness rising from the stomach, and, once in a long while, the painful but ecstatic transfiguration of his world.

In retrospect, the diagnosis was a blessing. Driving the bus had been a job without a future. It paid the bills, and that was all. On treatment for epilepsy, this career was not an option. Dave was forced to rethink. He spent time with the kids, with old friends, out in the countryside, fishing, reading. Alice returned to her previous work, as a primary school teacher in the village. Dave had a chance to study and to ponder. Eventually he decided to follow his heart and to do what he had wanted to, for ages. He began his training as a priest.

CHAPTER 6

Synapse

Dr Gelineau's Dream

ELISE

Dr Gelineau, Naval Surgeon of the Third Class, was still occupied in his surgery at midday, the Frenchman's sacred *heure de déjeuner*, seeing the last case of the morning. He had encountered only standard fare so far: a sailor with a fever, a pregnant woman anxious that she might be losing her baby, a leg wound refusing to heal, a ship's captain never likely to heed the advice – that he must abstain absolutely from alcohol – which he came to seek, for some reason, with such tiresome regularity. This last case was more puzzling, but Dr Gelineau's attention was starting to wander. Something was distracting him. This was his second long tour of duty in the Indian Ocean: he was thoroughly familiar by now with the sights and sounds of l'Ile de Mayotte. But the bells that rang in the distance, as if from the hills above the town, surely belonged elsewhere. With half his mind on his beguiling patient, he could not decide where.

She sat before him with her head half bowed, dark haired, dark skinned. She, also, reminded him of another time and place. Where was it? Ah – now he had it, the memory was of Elise, a woman he had treated on La Réunion two years ago. She had, quite literally, become crazed with grief after the death of her child. His patient raised her brown eyes and gave a small, shy, friendly smile. Gelineau started: surely this woman *was* Elise, but how did she come to be sitting with him, in his surgery, here and now? He had seen her last in St Paul's, the hospital he supervised on La Réunion, where he had prescribed quietness and rest – from, among other things, the inconsiderate demands of the naval officer who had fathered her child. She had no business to be here – but here she was, beyond all doubt.

Now she was beckoning him into the corridor which led from his surgery towards his own rooms, her hand, a moment later, resting unexpectedly, but pleasingly, on his, her Creole skin dark against his own. She drew him to her. He felt her warmth. As their bodies met he was filled with a fierce incongruous surge of joy, as if this unsought embrace were simultaneously homecoming, fulfilment and forgiveness. For a moment he was utterly at peace. Until the bells rang out again, louder this time, and Elise slipped from his arms. Dr Gelineau opened his eyes with great reluctance.

He was at home, in Paris. The bells tolled in the belfry of Saint-Sulpice. He could still smell the ocean, feel the pressure of Elise's body against his own, but the morning outside was dark and cold. He tried to move. For a few moments, he could not; he had often experienced this paralysis on waking from a dream, and it no longer disconcerted him. He waited until the paralysis passed – then swung himself from his bed, splashed his face with water, and dressed himself for a day of, no doubt, thoroughly civilised medicine.

THE BARREL MERCHANT

Among the patients who came to seek the help of Dr Gelineau during his later career in Paris as a 'specialist in nervous diseases', between 1878 and 1900, was a barrel merchant whom I shall call Monsieur Legrand. Legrand was 38 years old, muscular, a little overweight. He had two complaints, distinctive but presumably connected, as they had both interrupted his previously robust good health at about the same time, two years before. First, he had fallen prey to 'sleep attacks'. These irrresistible naps might overcome him as he spoke, in mid-sentence, or during a meal, the knife and fork dropping from his fingers, or at the theatre. He slept only briefly, for a few minutes, waking refreshed. But the attacks would take hold of him several times a day and were a source of increasing embarrassment. Once, in the Jardin des Plantes, around the monkeys' cage, a well-known 'rendezvous of the curious, the maids, the soldiers . . . he had fallen asleep with everyone round him laughing'.

Legrand's second symptom struck Dr Gelineau as more remarkable. Strong emotions and sudden changes of arousal – laughter, sadness, relief, anticipation – had begun to have a most peculiar effect on him. As he saw the joke, or realised that he had picked up a winning hand at the card table, he would weaken – his knees might buckle, his chin drop, his head nod: in a bad attack he would be unable to stay upright and crumple to the floor, or would slip helpless from his seat. Occasionally an attack of this second kind might end in a short period of sleep.

Gelineau was a well-trained, careful observer, fond of recording his observations. He published a thesis on the ailments he treated on l'Ile de Mayotte, wrote another on cases of chest pain at sea, and a play in verse, *After the Ball*. He was to publish nine further medical monographs during his years in Paris, and to contribute throughout his life to narrating the history of his home town, Blaye. So it was natural that he should wish to describe the intriguing case of Monsieur Legrand. The account appeared in the *Gazette des Hôpitaux de Paris* in 1880. Gelineau recognised Legrand's affliction as a distinctive disorder and coined a name for it: narcolepsy, from the Greek for 'caught by sleep'. He called the attacks of weakness induced by emotion 'astasia', though their technical description would later change to 'cataplexy', from the Greek for 'stupefied' or 'struck down with terror'. In 1881 Gelineau was able to gather together and publish descriptions of fourteen similar cases under the title *De la Narcolepsie*, establishing the existence of his mysterious syndrome, of sleep attacks and paralysis by emotion, firmly in the medical literature.

LUCY

Lucy felt her way along the passage which was almost pitch black, wet and dripping. It seemed endless. She must have been travelling down it for hours now, and it had become hard to tell whether danger lay ahead of or behind her. The walls of the tunnel were rough, but every so often her hand slipped over something soft and slimy, like a snail or a soft-bodied, naked slug. She shuddered at the thought. As a child she had dreamt of journeys like this, through a warren of tunnels, entered by a small door in the old-fashioned kitchen oven which she never saw opened during the day. At the very centre of the warren lived a witch. Lucy felt the same sense of dread now as she had then, and wished there was someone she could run to, sobbing with fear, as she had run to her parents from her nightmare. But this was no dream. Then, all of a sudden, things changed. The tunnel was warm, its lining soft around her, pulsing, oozing fluid, and it had grown lighter, glowing pink. Before she could inspect her new surroundings properly, the tunnel was tightening, clasping her, squeezing her, firmer and firmer until the breath was crushed from her and she was helpless, unable even to flinch from its suffocating pressure – then, at the moment when she thought she must surely lose consciousness, the tunnel relaxed, and Lucy drew breath with unspeakable relief. She heard a door open. Her brother put his head round it: 'You're dozing again, sleepyhead.' She raised her head from the desk on which she had slumped, and glanced at the crumpled school book open beneath her. She had been working on a story for her English class. Sleep must have overtaken her while she wrote: the essay tailed off with a single word, repeated several times in an increasingly wavering hand – 'helpless, helpless, help . . . '.

These irresistible naps had become a regular feature of Lucy's increasingly irregular existence. She was fifteen. Until a year ago her life had been smooth as a pebble. Lucy could count her troubles on the fingers of one hand. She had, in fact, often done so, lying in bed, in the arms of her own warmth, enjoying her drowsiness and the sense of impending sleep. But sleep, once friendly, had betrayed her. Now, when she meant to sit and work, sleep would creep up and take her by surprise. At night, when sleep was welcome, it would lull, but then desert her. She would have to get out of bed and read, or stroke her dog, who was at least pleased, if startled, to see her at 3 a.m. No one else seemed to be. And the sleeps which overtook her in the day were bizarre: she would be dreaming instantly, her perceptions as vivid, more vivid, than those of her waking life. Sometimes she seemed to catch a momentary glimpse of a dream unfolding independently, out of the corner of her eye, as if some part of her were dreaming secretly all the time. It was becoming increasingly hard to distinguish her recollections of dreams from memories of wakefulness: she had several times accused her brother of forgetting conversations with her which, on second thoughts, she must have dreamt. The boundaries between sleep and wakefulness were breaking down. Lucy, normally quick to question the sanity of others, wondered whether this was how madness began.

Last night something new had happened. She had been dreaming, again, of the tunnel. This time it ended in steps, steep, stony, spiralling down into icy coldness. She could hear the sound of water running, though the steps were dry. She was afraid, but excited too. There was something forbidden in this excitement: simultaneously she shrank back and wanted to press on. But then she had woken . . . or had she?

As she emerged from the dream she had felt a terrible weight on her chest, as if someone, or something, were pressing down on her. Try as she could, neither her arms nor her legs would move to help her push him – it – off her. This was the worst helplessness of all: unsure that she could breathe, sure she could not move, pinned to the bed by invisible bonds, or paralysed. After a few seconds the worst of her awful palpitating panic ebbed away. Whoever or whatever was holding her seemed to have no purpose beyond preventing her escape. Then, at last, the commands she was hurling mentally at her disobedient arms and legs began to do their work: she could move first a finger, then a hand, now her whole arm. Life flooded back into her limbs. She sat up and switched on the light, still breathless from the fright. The room was empty: she was alone.

That evening, when Lucy went down to eat with her brother and parents, Sam looked up as she came in: 'What's with you, Lu: you're asleep all the time. Who are you living it up with?' Lucy smiled, and was about to point out to her younger brother that she had spotted *him* furtively hand in hand with a girl on the way back from school, with pleasing anticipation of his embarrassment, when she felt her jaw sag and her knees buckle. Speechless, she just managed to manoeuvre herself on to a chair, as Sam contemplated her slow-motion collapse with a puzzled grin.

THE STUFF OF DREAMS

'Such stuff as dreams are made on . . .'
Shakespeare, *The Tempest*

Dr Gelineau would have been intrigued, or even gratified, by Lucy's confusing tale. He would have recognised at once that her sleep attacks belonged to his syndrome of narcolepsy, together with the 'astasia' or cataplexy which felled her gracefully at the moment she was about to play a winning card in the banter with her brother. These are the cardinal features of 'Gelineau's syndrome'. But unlike many sufferers with partial forms of this condition, Lucy has a full house of narcoleptic symptoms.

The dreams which come to greet her as she shuts her eyes to sleep are known as hypnagogic – literally 'sleep leading' – hallucinations; her unnerving experience at the hands of the incubus who sat upon her chest is sleep paralysis, quite common in the ordinary way of things, as Dr Gelineau knew from his own experience, but even more frequent in narcoleptics; her repetitive scrawl – 'helpless, helpless, help . . .' – as sleep overcame her is 'automatic behaviour', dopey, half-purposeful activity at a time of mounting drowsiness; her broken nights are, curiously, typical of narcolepsy, as narcoleptics have almost as much trouble remaining asleep as they do staying awake. This bizarre assortment of symptoms – sleep attacks, cataplexy, hallucinations, sleep paralysis, automatic behaviour, insomnia – seems to make no sense at all, until one finds one's bearings in the half-lit landscape of sleep. Work by the sleep physiologist Nathaniel Kleitman and his colleagues in Chicago during the 1950s revealed that sleep has a hidden structure. Part of the explanation for Lucy's predicament is to be found in their discoveries.

In the late 1920s a German psychiatrist, Hans Berger, had achieved a long-standing ambition. He had become fascinated by the goal of recording the electrical activity of the human brain, believing that this provided the material basis for the mind. In 1929 he published the first of several papers describing the human electroencephalogram, or EEG – the read-out of the brain's electrical activity recorded from the scalp. He and others, like Lord Adrian in England, soon recognised that its rhythms varied with our states of consciousness: beta rhythm, the electrical signature of attentive wakefulness, consists of irregular waves that repeat themselves 14–25 times each second; alpha waves are highly regular oscillations present over the back of the brain during quiet wakefulness, with the eyes closed, at 8–13 cycles/second; theta (4–7 cycles/second) and delta (less than 4 cycles/second) waves characterise states of deep sleep. Nathaniel Kleitman's work in Chicago in the 1950s explored the electrical patterns of sleep.

In the first half-hour or so of normal slumber the brain descends through a series of deepening stages into the comfortable mindlessness of 'slow wave sleep'. The brain's normally complex and rapid electrical activity is now dominated by regular waves occurring a few times every second – implying massive synchronisation in the activity of our neurons – and mental life slows almost to a halt. After half an hour or so of slow wave sleep, we reascend the ladder of sleep stages, and finally enter a 'paradoxical' state in which we are extremely inaccessible, floppy and difficult to rouse, but the brain's electrical activity is as rapid and changeful as in wakefulness. Occasional twitches, around the mouth or in the limbs, darting movements of the eyes, penile erection in men, betray the vivid inner life that accompanies this state: for at these times we dream. After a short spell of dream sleep, we descend the ladder once again.

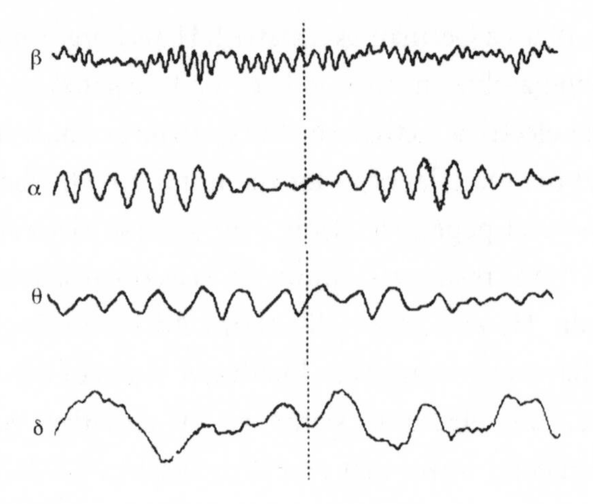

These four traces, each taken from a different subject, show the four rhythms most often encountered in the EEG over a two-second period: beta rhythm, at 14–25 cycles/second, is characteristic of the active wakeful brain; alpha, at 8–13 cycles/second, is seen during relaxed wakefulness, with the eyes closed; slower rhythms, theta, at 4–7 cycles/second, and delta, at less than 4 cycles/second, occur during normal sleep and in some kinds of coma.

In the course of a night this cycle repeats itself four or five times – with a subtle change in emphasis as the night proceeds: the bulk of slow wave sleep occurs in the first two cycles, while dreams are more abundant in the later part of the night. Waking at 2 a.m. we have to haul ourselves up from a still, deep well, while seven hours later we are much more likely to resurface with a dream in the mind's eye.

Work like Kleitman's pointed to the existence of three fundamental state of healthy consciousness: first, wakefulness; second, paradoxical or 'rapid eye movement' (REM) sleep; third, 'non-REM sleep', a category with four distinguishable stages: the first and second intervene between wakefulness and deep sleep, while in the third and fourth – slow wave sleep – the EEG is dominated by slow waves and we enjoy our deepest slumber. Our brains usually respect the boundaries between these states. Transitions between them are orderly, brief and uncomplicated – but not for Lucy.

Many of her symptoms are explained by an unusual fragility of the boundary between wakefulness and REM, states normally separated by a substantial period of non-REM sleep. Vivid images rush into her mind at the onset of sleep because her brain is already racing into REM while her mind is half awake. Paralysis at moments of emotion occurs when the floppiness or 'atonia' of REM – which normally prevents us from acting out our dreams – cuts awkwardly into wakefulness; paralysis on waking from a dream occurs when the REM atonia persists beyond the boundary of sleep. Automatic behaviour is another sign that transitions between sleep and waking are problematic: as when Lucy's pen marched on across the paper while her mind was already dozing. The fact that her nights are disturbed, that staying asleep is a problem too, suggests that maintaining *any* state of consciousness is a challenge for the narcoleptic brain.

We have known this much about narcolepsy – that it results from a failure to segregate the states of consciousness normally – for almost forty years. There were also clues, from rare cases in which narcolepsy appeared to develop following well-defined damage to the brain, that whatever causes the condition involves a region at the very centre of the brain, the hypothalamus. All this is true: but the crux of the matter turns out, remarkably, to lie in the absence of a single neurochemical. The missing ingredient is a neurotransmitter, required to convey a molecular message across the microscopic clefts, or synapses, between certain neurons critical to sleep.

DR GELINEAU'S CELLAR

In 1900, at the age of 72, Dr Gelineau retired to the country, to a castle, Sainte Luce La Tour in Blaye, where he had spent his boyhood. He devoted himself there not to medicine, but to viniculture – with fair success: his claret won a diploma of excellence in Amsterdam, and the great gold medal at the Paris Universal Exhibition. He must have had a discerning nose for the subtle chemistry of wines. It would not have surprised him to learn that sleep and waking also are governed by a balance of humours, playing on the brain. A neurochemical solution to the mystery of narcolepsy would have been congenial to Dr Gelineau.

The electrical signals transmitted by individual neurons down their axons are binary electrical messages that flick on or off like a switch. But, as Cajal had proposed, neurons are separated from one another by tiny gaps. These gaps are roughly the width of 200 hydrogen atoms – visible only at the immense magnifications provided by electron microscopes. They were christened 'synapses', from the Greek for 'clasp together', by Charles Sherrington, a British physiologist of the late nineteenth and early twentieth centuries who furnished these tiny clefts with a name some years before anyone had actually seen them. Cajal, architect of the neuron theory, convinced that neurons were separate from one another, had been fascinated by these, then invisible, points of contact. In his memoir he wrote of them passionately, echoing Sherrington's choice of words: 'What mysterious forces finally establish those protoplasmic kisses, the intercellular articulations, which seem to constitute the final ecstasy of an epic love story?'

While the signals travelling down neurons are electrical, the synaptic gaps between them are bridged by chemistry: a liquid

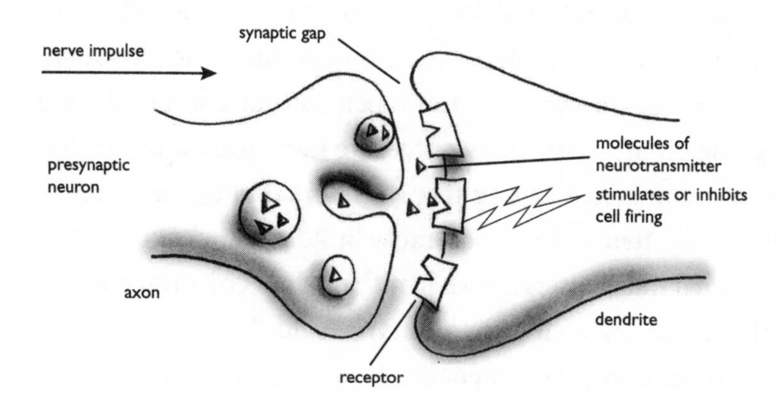

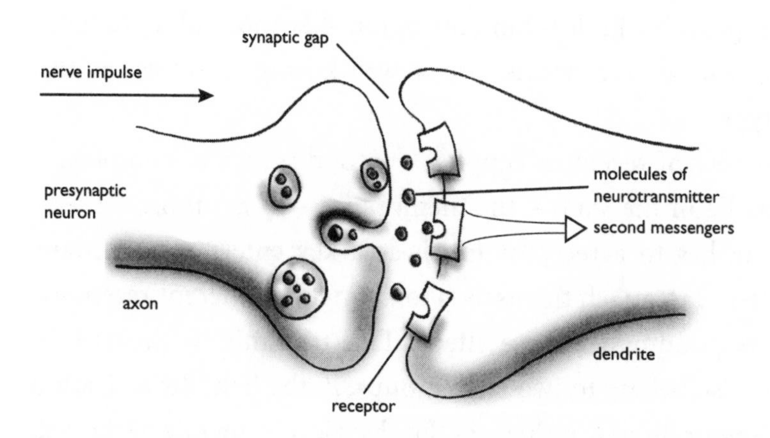

6. The Synapse
Neurotransmitters are released from neurons into the synapse between one neuron and the next. They lock into ëreceptorš on the far side of the synapse, where they have two main effects: they may directly alter the likelihood that the neuron they lock onto will fire (increasing or decreasing the chances), or they can influence biochemical processes occurring within the cell, including gene activation, through the production of chemical 'second messengers'.

'neurotransmitter' is released by the terminal of one cell, to be picked up by a 'receptor' on the next. Although this principle sounds simple, the process involves several dimensions of complexity. For one thing, the brain's neurotransmitters are numerous, around 30 at the present count. They belong to two main families: small molecules, which are often amino acids, those building blocks of proteins we encountered in Chapters 2 and 3, or their close chemical relatives – like glutamate, gamma-amino-butyric acid (better known as GABA), acetylcholine, noradrenaline, dopamine, histamine and serotonin, and larger molecules which are essentially small proteins known as peptides. Endorphins, the opium-like molecules produced naturally within our brains, are a well-known example of this second type.

The variety of receptors supplies a second source of complexity. Receptors lie in the walls – the membranes – of neurons. A transmitter attaches to a receptor much as a key enters a lock. Many transmitters can attach themselves to a range of different receptors, with correspondingly diverse effects. Like transmitters themselves, receptors also belong to two large groups. In the first, the activation of the receptor directly influences the electrical firing rate of the cell, by allowing charged particles of sodium, potassium, calcium and chlorine to flow in or out; in the second group, receptors generate a series of chemical 'second messengers' within the neuron. These messengers can trigger processes ranging from gene activation and protein production to indirect effects on the cell's firing rate.

Henry Dale, one of the pioneers in the study of neurotransmission, sugggested that a single neuron always and only released a single transmitter at its synapses, but it is now clear that co-release of two transmitters, often a small molecule and a peptide, is common. As a neuron can both make and receive several thousand synapses, some inhibitory, some excitatory, some directly coupled to its firing rate, others acting via second messengers, there is plenty of scope for interactions of effect. The play of transmitters over a neuron's numerous synaptic connections with other cells determines its ultimate behaviour.

How do the workings of the synapse help to explain the plight of Lucy and Legrand? Since the work of the Belgian physiologist Frederick Bremer in the 1930s, and of his student Giuseppe Moruzzi in the 1940s, we have known that the cycle of sleep and waking is controlled by clusters of neurons in the 'stem' of the brain – the region connecting the spinal cord below and the hemispheres above. These neurons release their neurotransmitters throughout the brain, acting on the cells they contact through second messengers. Collectively, they create an 'activating system' which controls our states of awareness.

Four neurochemicals in particular, acetylcholine, serotonin, histamine and noradrenaline, modulate conscious states. The activity of all four is high in wakefulness and declines in slow wave sleep. During REM sleep, clusters of neurons transmitting acetylcholine become highly active while cells producing the other three transmitters fall absolutely quiet. REM periods end as this balance reverses. But these fundamental mechanisms are intact in Lucy and Legrand. The chemical key to their disorder was discovered quite by chance in the late 1990s.

THE SILENT NOTE

It would be hard to improve on Dr Gelineau's clinical description of the symptoms of narcolepsy. But he could not know what caused this strange constellation of sleep attacks, paralysis by emotion and incontinent dreaming. Almost 120 years were to pass before the basic sciences of neurochemistry and genetics became able to reveal the subtle defect. The discovery eventually came from a completely unexpected direction – a biochemical fishing expedition, trawling for novel chemicals in a small but crucial region of the brain, the hypothalamus.

This small cluster of cells at the base of the brain, only the size of a small peanut, is the principal crossroads between the brain and the wider internal functioning of the body. It monitors, for example, the blood sugar and the salts in the blood. As these change, the hypothalamus responds by arousing or suppressing hunger and thirst, and by varying the release of hormones, from the pituitary gland which lies just below it. These, in turn, adjust our metabolism and urine output to maintain a relatively constant milieu within the body. Among the many other functions of the hypothalamus, it helps to regulate the cycle of sleep and waking.

In 1997 a study of the genes that are particularly active in the hypothalamus revealed a number of novel proteins. One of these turned out to stimulate appetite in rats and was christened 'orexin'; another group of scientists noticed the resemblance of the same substance to the hormone 'secretin': as it was found in the hypothalamus they called it 'hypocretin'. A search of a bank of 'orphan receptors' – brain molecules resembling known receptors, identified biochemically but not yet linked to any known transmitter – revealed one which bound hypocretin tightly. This suggested that this newly discovered chemical was very likely to be a hypothalamic neurotransmitter.

The discoveries which followed tumbled over one another very rapidly, reflecting both the sophistication of modern molecular medicine and the fruits of the preceding fifty years of research on narcolepsy in man and animals. First, it emerged that dogs – Dobermans, Labradors and dachshunds – with inherited narcolepsy, which had been intensively studied for many years at Stanford University in California, had mutations in the gene for the hypocretin receptor. These mutations abolished its function. The serendipitous discovery of this new neurotransmitter system was leading to unexpectedly interesting results. Next, experimentation strengthened the evidence linking narcolepsy to hypocretin. It is now possible to delete genes deliberately from animals to learn more about their function. A strain of mice, in which the gene for hypocretin had been artificially deleted using the tools of molecular genetics, displayed curious periods of 'behavioural arrest'. These turned out to be intrusions of REM sleep into wakefulness: these mice had cataplexy.

This work showed that genetic narcolepsy can be caused by mutations in the genes for hypocretin or its receptor. But human narcolepsy is not straightforwardly inherited, even though the risk of developing the disorder is raised if a close relative suffers from it. Could it be, nonetheless, that hypocretin is the key to human narcolepsy? In 2000 it was first reported that human sufferers from narcolepsy have unmeasurably low levels of hypocretin in their spinal fluid. Examination of a small series of brains donated by patients with narcolepsy revealed the selective, more or less complete, loss of hypocretin-producing cells from the hypothalamus. These cells are few at the best of times, just a few thousand neurons among the hundred thousand million in our brains. Their loss, and with it the loss of hypocretin from the brain, appears to be the root cause of human narcolepsy.

How can we link the disappearance of this small population of neurons to the complex predicament of Lucy and Legrand? Why does the absence of a single scarce neurochemical allow the partitions between conscious states to break down? Hypocretin is conveyed from the hypothalamus widely through the brain, but especially to the groups of neurons in the brain stem which regulate sleep and waking. It stabilises their activity, helping us to resist sleep during wakefulness, to postpone our dreaming once asleep, and, more generally, to consolidate our conscious states. Why should *just* those neurons which produce hypocretin disappear selectively from the brain in narcolepsy? The current consensus is that these cells are lost as the result of a self-directed attack mounted by the sufferer's immune system on his own cells, but the positive evidence is slim so far. Ground-breaking work in science tends to pose as many questions as it solves.

This story of discovery should have a practical outcome some time in the next few years. Now that the cause of narcolepsy has been identified as the loss of a single neurotransmitter, the hunt is on for a preparation that can be used to treat the disorder, a form of hypocretin that can regulate arousal, just as insulin is used in diabetes to regulate the blood sugar. Such treatment is badly needed: as Lucy and Legrand both discovered to their cost, irresistible sleepiness recurring throughout the day, and paralysis by strong emotion, are not in the least funny for the sufferer – even if unkind bystanders can sometimes find them so.

It has taken a century to move from the clinical description of the narcoleptic syndrome to the molecular description of its cause. The story of narcolepsy is the history of neuroscience in miniature. Understanding it required the definition of the neuron by Cajal; the discovery of the synapse by Sherrington and his followers; the description of the neural basis of our states of consciousness begun by Hans Berger, the German psychiatrist who first recorded the electrical rhythms of the human brain, and Frederic Bremer, the physiologist who pointed to the regions in the brain stem that control them; and, finally, most recently, the biochemical exploration of the numerous transmitters that carry messages across synaptic clefts. The scientific story is remarkable, but it was Dr Gelineau's simple but memorable clinical description that provided the point of departure for the ensuing odyssey. He would, surely, have loved to hear the tale.

CHAPTER 7

Neural Network

The Sense of Pre-existence

FROM CELLS TO SYSTEMS

By now you have encountered the basic building blocks of the brain, its cells and synapses, and its basic ways of working: electrical signalling along the neuron, chemical signalling across the synaptic gap. These cells and synapses are prodigiously numerous: there are thought to be around a hundred thousand million nerve cells in the human brain. Surely there must be some organising principles, between microscopic neuron and macroscopic brain, which help to explain how they go about their business of enabling our behaviour and experience.

As it turns out, there are indeed several levels of organisation beyond the neuron – local circuits consisting of a few hundreds or thousands of neurons in highly compact columns or systems of cells; more extensive, but equally regular arrangements within brain regions that are large enough to see with the naked eye, and 'distributed' networks that interlink several parts of the brain in the service of a shared goal, like playing a volley at tennis or remembering your last evening out with a friend. Correspondingly, besides the incessant crackling of signalling in individual neurons, the brain is alive with rhythms of activity that express these neurons' summed, coherent working in its networks.

You remember that evening now? Memory, as it happens, is not a bad place to start getting to grips with the networks in the brain. I shall begin with one of its more familiar misbehaviours.

DÉJÀ VU

Déjà vu is the disconcerting sense that our current experience echoes some ill-defined, terminally elusive, past experience. Most of us have experienced it now and then. Occasionally it can take on pathological proportions.

One day when she was in her mid-twenties, Shona, whom I met years later, found herself in a disconcerting state of mind. Rising from bed that morning, washing, dressing, preparing breakfast, leaving for work, she felt she was continuously 'acting in a film that she had seen before'. All these humdrum activities really were, of course, truly familiar – but somehow things were different that morning: she felt she had lived *just these* same moments, *just this* same day before. She was mysteriously caught up in a repeat performance, point for point: throughout the day she had the sense of knowing precisely what would happen next. She wondered if she was over-tired or sickening for something. But despite a rest and some aspirin, she was unable to shake off the weird, tormenting sense of re-enactment.

If you have experienced *déjà vu* once – and you probably have – you will have had it several times. Reports of *déjà vu* rise with years of education, as they do with social class. Regular travellers, and those who remember their dreams, are more prone to the experience than home-loving, dreamless folk. The frequency of episodes peaks in the third decade of life and declines thereafter. It occurs more commonly in the eve than the morn, and is usually all over within seconds. It can be triggered by fatigue and stress.

It is, even in its ordinary, momentary form, an arresting – indeed memorable – experience. It embraces a contradiction. On the one hand it impresses us with its air of significance. The act of recognition during *déjà vu* is more intense than usual, as if the present experience has made contact with 'the very back scene of the theatre of things', G.K. Chesterton's haunting description of his earliest memory. But, at the same time it is fragile: for, in the words of a nineteenth-century sufferer, we half realise that the recognition 'is fictitious and the state abnormal'. Experiences like *déjà vu* lie at the fringe of consciousness, capturing our attention just at the moment of their disappearance. It may be precisely because we are unable to get a proper hold of them that they seem to open a door, for an instant, to another universe of possibilities, abiding but concealed, beguiling but unattainable.

Shona's experience was clearly different, more disturbing and prolonged than the everyday varieties of *déjà vu*. A contemporary expert on the experience, the British psychologist Martin Conway, has named this kind of experience '*déjà vécu*', literally 'already lived through', to underline its key feature, the conviction that one has lived through exactly these events before. This can be strong enough to act upon: Shona rebuked a friend who embarked (for the first time) on 'the same conversation all over again', and unplugged the radio because the news was always old.

After three or four days of unremitting *déjà vu*, Shona sought help. By then her experiences had become even stranger. The left side of her face was tingling on and off. At times she had a weird sense that she had left her body and was looking down on it from above; at others, she felt that she was somehow being compelled to do things, that her actions were no longer her own. She was admitted, initially, to a psychiatric hospital. A detailed assessment there unearthed the fact that Shona had suffered from epilepsy as a child. An astute psychiatrist wondered whether these strange experiences might also be epileptic. She was transferred to a neurology ward. A brain wave recording confirmed that there was indeed continuous epileptic activity over a region on the right side of her brain. After a couple of days of treatment with antiepileptic drugs Shona was greatly delighted to find the world restored to its only roughly familiar, obligingly unpredictable old self.

Déjà vu of every kind is a wrinkle in the normally smooth processes of recognition. Recognition is a species of memory. Learning more about how the brain enables memory should help to illuminate the cause of Shona's plight. Doing so will introduce us to some of its more complex circuitry, but let's begin the story simply.

SIMPLE CIRCUITS

In our ascent through the levels of the nervous system, from the atoms that compose us, through the constituents of our cells, to the neurons championed by Cajal and the synapses christened by Sherrington, we have not stopped for long to consider what brains are for. They are, of course, as I have briefly mentioned, for communication and control. Bodies are great assemblages of cells, bundled up into organs with specialised roles. The kidneys filter the blood that the heart pumps and the lungs oxygenate. These organs are splendidly equipped for their tasks. But all this marvellous workmanship would be to no purpose were it not for a command system that organises the activities of the whole organism, on the basis of its needs, in the light of what is happening now and what has happened in the past.

To see how nervous systems do their job, it's a help to begin by considering either very simple ones, or very simple parts of complex ones. Charles Sherrington, the great British physiologist of the nervous system, used the second strategy at the close of the nineteenth century, when he studied reflexes, like the knee jerk. When the knee is tapped over the tendon that attaches the 'quads' to the lower leg, the muscle rapidly contracts, raising the foot in the air. Sherrington showed that this 'atom' of behaviour depends on a network of neurons, but an extremely basic one – a 'sensory' neuron, originating within the muscle, detects and signals the stretch of the tendon by the tap. It runs back to the spinal cord, where it makes a synapse on to a 'motor' neuron that sends its own axon back to the muscle, causing it to contract.

Similar reflexes, for example the 'gill withdrawal reflex', have been studied in intimate detail in simple organisms like the California sea slug, *Aplysia californica.* The relevance of this work for us is that these reflexes, though predictable and automatic, by and large, are not unalterable, either in us or in *Aplysia.* With simple repetition the reflex gradually diminishes or 'habituates'; with 'facilitation' the reflex can be enhanced. The reflex, in other words, is plastic: its current behaviour reflects its past experience. In these extremely simple modifications of behaviour by experience in the simplest of neuronal networks lie the pregnant seeds of human memory. In fact, the ability to adapt to changing circumstances, plasticity, is ubiquitous in the body. The athlete's slow pulse, the leathery skin on hard-used soles, the drinker's remarkable capacity for booze, all reflect the malleability of our tissues. But neurons have developed this capacity in spades. The place in the nervous system where experience can make its mark most readily, its most impressionable corner, is the synapse.

Research involving *Aplysia* and other simple organisms has allowed close scrutiny of the steps by which the effects of experience are inscribed on the brain. The synapse turns out to be highly dynamic. In the short term – over seconds to minutes – the effects of activity in one neuron on the behaviour of another can change rapidly, with variation in the amount of neurotransmitter released between them. A reduction in transmitter release underlies the decline seen in *Aplysia*'s withdrawal reflex with repetition. Over longer periods, minutes to days, synapses visibly enlarge or shrink as the patterns of activity that play across them change: both the incoming axon and the dendrites that receive its news will branch, and expand if the signalling between them is intense – or shrink and retract if it subsides. The resulting alliances are specific, building links between neurons that are activated together by particular events or that collaborate in particular tasks. The eventual outcome is that 'cells that fire together wire together'. In the short term, the modifications at the synapse are accomplished by easily reversed chemical reactions. The growth and shrinkage of axon terminals and dendritic trees require more time-consuming processes, of gene activation and protein synthesis. But, over time, these structural changes at the synapse create the efficient, well-worn routes through networks of neurons that ultimately underpin our habits, skills and memories.

SLEEPING AND FORGETTING

Before we trace the path by which complex networks developed from the simple ones that suffice for a sea slug's needs, I will introduce you to a second quirk of memory – more or less the flip side of Shona's difficulty.

Jed, a middle-aged academic contacted a colleague of mine, via the web, after coming to the conclusion that he was suffering from a form of amnesia that we were studying. As it turned out, he was absolutely right. He had kept a detailed diary of events from which I have reconstructed his thoughts and feelings when he woke, one morning, in a place that he found alarmingly unfamiliar –

> I'm in trouble again. I woke half an hour ago in a room I didn't recognise. The air feels different, the sounds from the street are foreign, a suitcase lies beside my bed. The room is incongruously tall and narrow, high windows, bright sunshine filtering through silk curtains: Where on earth am I?
>
> In a state of some bemusement I make my way to the window, draw the curtains, look out. A glance is enough to tell me that I'm in Italy! What am I doing here? There is some kind of parade in the street below, guys in medieval costumes carrying tasselled banners, girls with drums, small pipes, mothers with babies in their arms, cars hooting round the corner. You can tell the day will be baking hot. A Scottie dog glances up at me thoughtfully from the sidewalk, cocks his head. If he is puzzled, well, so am I. I look back into the room to get some clues. The clothes draped over my chair are northern, wintry. They don't belong to this hot Mediterranean city. A summer suit, freshly minted from the cleaners, lies at the top of my case. So: I have travelled recently. I rifle through my pocket, find some tickets – 30 April flight from New York, a bus ticket to Sienna. So it's May day. I'm in Tuscany. And I don't know why . . .
>
> I'm beginning to remember now. I can peer through a few clear islands in a misted pane. Others are clearing as I look. I'm here to

> lecture. I've searched 'My Recent Documents'. There's a Powerpoint – 'Sienna, 2 May: 'Dark Matter in the Milky Way', a popular talk I've given a few times now. I remember that Barbara flew off to London a week ago, and we're meeting up in Rome, 10 May. I think I'm meant to be going into the mountains before then, to see the particle detector. But I still can't remember who invited me here, or what happened back home the days before I came. What worries me most is that I'm sure this isn't the first time that I've woken in this state. I can remember not remembering before.

Déjà vu is a brief excess of memory. Transient amnesia, the temporary loss of access to old memories, or temporary inability to form new ones, is its negative counterpart. We all encounter its minor forms from time to time. Dan Schacter, the American psychologist, includes memory blocking among his Seven Deadly Sins of Memory. These comprise two other types of forgetting – due to fragility of memories and absent-mindedness – three varieties of memory distortion – misattribution, suggestibility and bias – and finally persistence, the intrusion of unhappy memories we would rather be without. The frequency of the everyday memory symptoms that result from these 'sins' rises relentlessly with increasing age – as the frequency of *déjà vu* gradually declines.

Normally we run up against these roadblocks one at a time: we lose a name, or can't recall that day trip to a lake we're told we took five years ago. Jed's amnesia was a more widespread traffic problem, ranging across the entire extent of his recent memory. But his memories were not irretrievably lost. His gradual reorientation, as his memories swam back into view, shows they had only temporarily slipped from view. Why did he wake and forget?

Jed's predicament was unusual, but not unique. Alarming, but usually solitary periods of amnesia, lasting for three or four hours, occur out of the blue in a condition called 'Transient Global Amnesia'. This affects people in their mid to late life at a rate of about one per 10,000 per year. Episodes are often triggered by excitement or stress: a swim in a cold sea, a quarrel, an illicit sexual encounter. Abruptly, sufferers find they are unable to remember. As they cannot recall recent events, they lose their bearings in time and place; as they cannot remember the answers given to their insistent questioning – 'Where are we? What have we been doing?' – they remain rudderless for the period of the attack, of which they have no subsequent recall. Attacks like Jed's, brief, recurring episodes of amnesia on awakening, are a variation on the theme, a recognisable sub-type within the wider family of transient amnesias. His forgetfulness on waking, like Shona's *déjà vu*, was due to local epileptic activity scrambling the neural processes required to retrieve memories. Blows to the head, attacks of migraine, certain drugs, 'mini-strokes' impairing the brain's blood supply can all produce the same or similar effects.

Shona could not stop herself remembering things that hadn't, in fact, ever happened before. Jed couldn't remember the things that had. How do those extremely simple circuits that underpin reflexes from sea slug to man organise themselves into the networks that normally allow for the richly coloured recollection of times past?

THE ENCHANTED LOOM

The brain is waking . . . . Swiftly the head-mass becomes an enchanted loom where millions of flashing shuttles weave a dissolving pattern, always a meaningful pattern though never an abiding one . . . .
Sir Charles Sherrington, *Man on His Nature*

The transformation of simple into complex nervous systems does not involve a change in kind. The small, much-studied, nervous systems of roundworm and sea slug are built from just the same basic elements – neuron and synapse – as the magnificent brain of man, and employ the same key chemical processes. The differences lie in numbers, a host of subtle but significant variations on the basic neuronal theme and the gradual evolution of elaborate networks of neurons.

The scale of the change in numbers is almost inconceivable. *Caenorhabditis elegans* is a roundworm extensively studied by geneticists and neurobiologists over the past fifty years. It has exactly 959 cells, of which 302 are neurons. *Aplysia*, the sea slug, has around 20,000 neurons in its more elaborate nervous system. You and I have, very roughly, a hundred thousand million neurons in our brains, 1,000 million times as many as *C. elegans*, five million times as many as *Aplysia*, twenty times as many as there are human beings on the earth. Even these enormous neuronal numbers are dwarfed by the numbers of connections they permit: a single neuron may make up to 1,000 synapses on to other cells and receive 200,000. During the first six months after birth, when neurons are busily forming the networks that will provide the basis for behaviour and experience throughout the rest of life, 100,000 synapses form every second in the area of the human brain devoted to vision alone.

As the numbers of neurons and synapses increased in the evolving nervous systems of our ancestors, so these basic elements diversified: not by changing their fundamental nature, but by a series of subtle modifications that gradually created the huge variety on show in the human brain. Within its tangled wood, explorers like Golgi and Cajal discerned a wealth of neuronal varieties, and went on to name them – pyramidal, stellate, basket, fusiform and chandelier cells, to mention a few. The synaptic meeting places of these neurons are just as various: some are close to the 'axon hillock' where the axon's outgoing electrical signal is triggered; others, on the finer twigs of the dendritic tree, exert a less commanding but still appreciable influence. Within the synapse itself, variety was gradually generated by increasing the range of neurotransmitters used to signal from one cell to the next, and by multiplying the receptors that provide the locks in which the transmitters turn their chemical keys. Serotonin, for example, is a neurotransmitter found in the simple nervous system of the sea slug *Aplysia*: it has risen to recent fame as the chemical target for Prozac, the most famous of antidepressants. Serotonin acts on around thirty subtly different receptor types in the human brain, falling into seven main family groups. These numerous sub-types allow a single neurotransmitter to exert subtly varying effects at different sites in the brain.

A third process has also been at work in the development of complex brains. The vast numbers of neurons, in their numerous varieties, are linked into networks – local, extended and distributed. The region of the brain most closely linked to memory – as we shall see – provides a good example. The hippocampus is a curving structure that resembles a sea-horse (the source of its name) in cross-section. Tucked into the hidden inner surface of the brain, about four centimetres long and one centimetre wide (see Figure 7) the

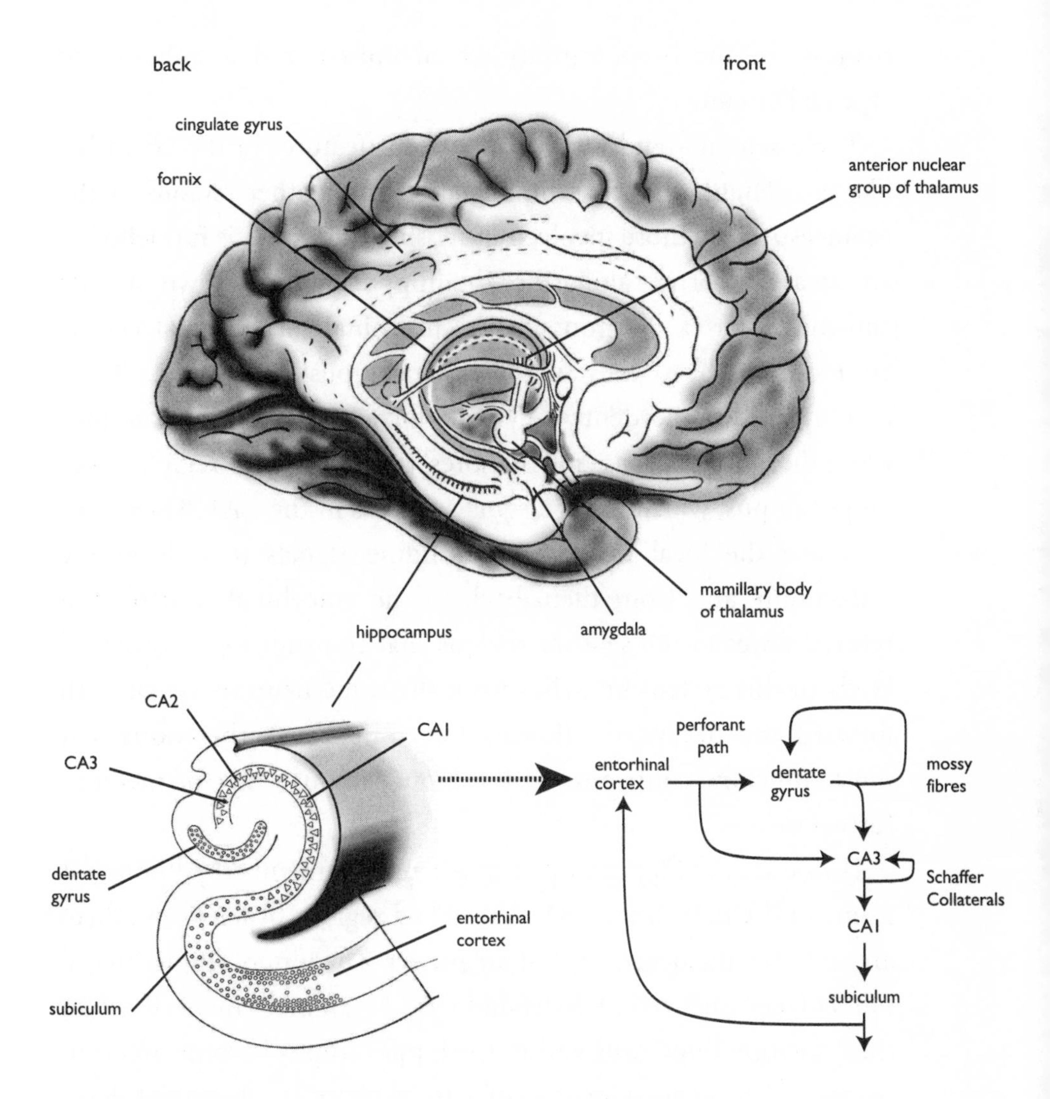

Neural Networks: The Limbic System and the Hippocampus
This view shows the inner surface of one of the brain's two hemispheres. The limbic system is highlighted. Damage to hippocampus, fornix or parts of the thalamus - a 'network of networks' - can cause permanent amnesia. Information funnels from widespread brain regions through the seahorse-shaped hippocampus (shown in cross section at bottom left). The figure at bottom right illustrates the network of interconnection to, from and within the hippocampus (CA1-CA3), as described in the text. This network within the hippocampus is thought to create an index that allows the recollection of recent experiences.

neurons of the hippocampus are organised into a million-fold repeated circuit.

The essential – and much-simplified – features of the circuit go like this. Highly processed information from other regions of the brain, especially those involved in sensing the world, is funnelled to an area of cortex alongside the hippocampus known as the entorhinal cortex. Neurons in the entorhinal cortex signal via the 'perforant path' to cells in the hippocampus itself (in its 'CA3' region) and in the dentate gyrus, which hugs the hippocampus. From the dentate gyrus, 'mossy fibres' run to the CA3 region of the hippocampus, which in turn signals to cells in the CA1. These cells complete the local network, by sending signals to cells in the 'subiculum' and from there back to the entorhinal, which itself returns fibres to the sensory regions that transmit to it. At several levels in this system branches from the same neurons signal both forward and backward, allowing for complex reverberations that must somehow contribute to the hippocampus's task of memory formation.

How does the hippocampus enable us to remember? Although it is one of the most intensively researched regions of the brain, there are still only the beginnings of an answer. The hippocampus 'maps' our current and recent surroundings. As animals move through their surroundings, cells within the hippocampus become active in sequence. The same sequences of activity recur in subsequent sleep, as if the animal were rehearsing, and perhaps consolidating, the memory of its recent experiences. Disrupting the function of the hippocampus during this early period disrupts memory formation. The connections between hippocampal neurons, like the maps that they create, are highly modifiable, as they must be if they are to keep track of changing experiences. The mechanisms of the process of 'long-term potentiation' that is now thought to underlie the capacity for learning in the hippocampus are understood in some detail, but there is disagreement about many other aspects of hippocampal function. The 'standard' theory of memory formation holds that, over days or weeks, memories destined to endure are transferred from their temporary index card within the hippocampus to a more robust, but more slowly formed, record elsewhere in the brain. The underlying notion here is that it is an advantage to have two systems for learning about recent events in the brain: a fast system, taking quick notes that are easily overwritten, and a slower system, creating a more selective and durable record. Although it has inspired much computational modelling of human memory, this theory is now under fire, as there is accumulating evidence that the rich recollection of individual episodes in fact requires a contribution from the hippocampus all the way through our lives.

Just as neurons are linked anatomically into networks, so their individual electrical activities combine to produce rhythmic patterns. This happens when their discharges synchronise. Too much synchronisation is not a good thing for the working brain – it characterises sleep and epilepsy, as we have seen with Dave and Lucy – but the right amount of delicately regulated synchronisation is essential for normal brain function. There is a musical analogy. Playing a symphony requires a harmonious combination of the various strings, wind, brass and percussion: if everyone played at random, or if everyone in the orchestra played the same notes simultaneously, the result would be either chaos or monotony. Like an orchestra, the brain has its own harmony of parts: a range of interweaving patterns of rhythmic discharge, at frequencies from a few per second to hundreds per second, each composed of the firing of thousands of cells, varying with our activities, is vital to the normal functioning of the hippocampus.

The hippocampus provides an example of both a local and an extended network. A single instance of the circuit I described involves just a few hundred cells, and is visible only beneath a powerful microscope. Extended many hundreds of thousand times along the length of the hippocampus, these circuits' hippocampal home is visible to the unaided eye. But memory, of course, requires more than a working hippocampus. Like the other major functions of the brain – sensing, communicating, acting – it depends on a network of networks, distributed round the brain.

AT THE LIMIT

> the smell and taste of things remain poised a long time, like souls,
> ready to remind us, waiting and hoping for their moment . . . and
> bear unfaltering, in the tiny and almost impalpable drop
> of their essence, the vast structure of recollection.
>
> Proust, *A La Recherche du Temps Perdu*

Memory is ubiquitous in the brain because synapses are found everywhere, and everywhere they are plastic. It is no surprise therefore that memory comes in many shapes and sizes: from the waning of an overworked reflex, through the capacity to learn a skill or recognise a face, to the ability to recollect past experience. This last is the type of memory that generally comes to mind first: memory proper, Proustian memory. It turns out that this species of memory, recollection itself, depends on a distinctive 'network of networks' identified and christened by the nineteenth-century

French neurologist Paul Broca well before its role in our mental life became clear.

Broca is most famous for his work on the embodiment of language in the brain. By studying the brains of patients with language disorders after their death he had concluded, in the 1860s, that damage to the front of the brain's left hemisphere caused the loss of fluent speech. He proposed that language was mainly a function of the left side of the brain. This was a fundamental insight, hinting at a deep asymmetry in the functioning of the human brain. 'Broca's area' is now known to be one of several areas in the brain, mostly left-sided, that are linked in a 'language network'.

Broca was a productive neuroanatomist as well as an astute clinical observer. During his explorations of the brain he identified another set of regions that runs around the inner edge of the cerebral hemispheres, creating a kind of inner lip. He christened these the 'limbic lobe', after the Latin, *limbus*, meaning border or edge. Nowadays the limbic system includes the hippocampus; the amygdala, a nut-shaped structure named from the Latin for 'almond' that lies alongside the head of the hippocampus; and the cingulate gyrus, from *cingulum*, Latin for 'belt', a curved expanse of cerebral cortex that completes the upper part of a rough circle which has the hippocampus in its lower half. They are linked in the 'circuit of Papez', which also involves the fornix, a fibre bundle running from the hippocampus to regions in the thalamus, a crucial relay station for neural signals of all kinds at the centre of the brain. Within each of these areas, neurons are organised in local networks – the circuit of Papez is therefore a network of networks.

Broca speculated that these linked limbic areas must share a common role. So it has proved. The role that has emerged would have fascinated Proust – or indeed anyone who has relished the recollection of emotion in tranquillity. The limbic lobe contains the major cortical centres serving the sense of smell. These lie alongside neurons concerned with emotion, found particularly in the amygdala, and others concerned particularly with memory, in and around the hippocampus. Their interconnected functions come to light in ordinary experience, and in disorders of the brain.

Scents are notoriously evocative. A lover's perfume encountered years after the end of the affair, like the aroma of Proust's famous madeleine, can suddenly awaken a lost world of experience. The resulting recollection is 'multimodal', combining memories of sights, sounds and touch, and always has its fringe – or limbus – of emotion. Smell, memory and emotion coalesce in our experience, as they do in the limbic lobe. The limbic lobe is also a common source of epileptic seizures. These can activate the limbic system, giving rise to intense emotion, most often fear; to strong, 'indescribable', smells, perceptible only to the sufferer and to disturbances of memory, both *déjà vu* and transient amnesia – like Shona's and Jed's.

Lasting damage to the limbic system, correspondingly, tends to injure memory. One of the most famous patients in the history of neuropsychology, a Canadian known by his initials HM, suffered from epilepsy arising from the limbic lobe. His hippocampi were surgically removed from both sides of the brain, in an attempt to improve his epilepsy, before their importance was fully appreciated. His seizures became less disabling, but his experience thereafter was locked into a perpetual present, lasting for just a minute or so: his ability to form new, long term, conscious memories of events had vanished beneath the surgeon's knife. A more familiar form of amnesia resulting from damage to this region of the brain is increasingly widespread today: the first pathological signs of Alzheimer's disease appear in the entorhinal cortex, the brain region next door to the hippocampus. The first symptom of Alzheimer's disease, correspondingly, is an inability to recollect recent experience. Mischief elsewhere in the circuit of Papez also gives rise to amnesia: alcoholics who give up food in favour of the bottle are at risk of vitamin B1 (thiamine) deficiency, which damages the circuit in the thalamus, causing the Wernicke-Korsakoff syndrome. This leads initially to confusion, unsteadiness and disorders of eye movement. Untreated, the confusion is replaced by permanent memory loss.

Other varieties of memory, in case you were wondering, depend on other regions of the brain. Short term, or 'working' memory, the ability to hold information in mind while we rehearse or work with it, involves sustained activity in the parts of the brain that normally deal with that kind of information – for example the language network for words – combined with activity in 'executive' regions, particularly at the front of the brain, that allocate attention. 'Procedural' memories, for how to do things, like riding a bike, are, reasonably enough, especially dependent on areas linked to the control of movement, like the cerebellum, the 'little brain' tacked in behind the hemispheres, and the basal ganglia, which we encountered with Charley in Chapter 2. Brain regions concerned with perception, like the visual system, contribute to forming and storing perceptual memories, for how things look or sound.

Jed's temporary amnesia resulted from the brief inactivation of the limbic network required for retrieving the memory of recent events. How did Shona's excess of memory arise? Some light has been shed on her perplexing experience by a recently observed subtlety of memory processing in the limbic system.

'YOU HAVE BEEN MINE BEFORE'

The experience of *déjà vu* has been described repeatedly by novelists and poets, seduced by its numinous quality, its hint of reincarnation, the intimation that it might reveal the 'secret of a life'. I borrowed the title of this chapter from Sir Walter Scott:

> . . . that yesterday at dinner time I was strangely haunted by what I would call the sense of pre-existence – . . . a confused idea that nothing that passed was said for the first time, that the same topics have been discussed, and the same persons had stated the same opinions on the same subjects.

Something about the nineteenth century seems to have been congenial to *déjà vu*. Dickens describes it evocatively in *David Copperfield*:

> We have all some experience of a feeling which comes over us occasionally, of what we are saying or doing having been said or done before, in a remote time – of our having been surrounded, dim ages ago, by the same faces, objects and circumstances – of our knowing perfectly what will be said next, as if we suddenly remembered it.

Soon afterwards, Dante Gabriel Rossetti transformed the feeling into haunting verse:

> I have been here before,
> But when or how I cannot tell:
> I know the grass beyond the door,

The sweet keen smell,
The sighing sound, the lights around the shore.

You have been mine before . . .

'Sudden Light' (1870)

Reincarnation may strike you as an unsatisfactory explanation of the experience. If so, contemporary psychology offers three broad alternatives. The first idea is that the sense of mysterious repetition at the core of *déjà vu* does indeed reflect an unusual recurrence, but the recurrence lies in the brain: the single smooth stream of brain activity that normally underlies perception is somehow split in two, so that one stream can be experienced as an echo of the first. The second suggestion is that something is indeed familiar in *déjà vu*, but the memory concerned has become inaccessible, creating the intense but apparently groundless sense that 'you have been here before'. The third theory is the most topical. It stems from the discovery that part of the memory network in the limbic lobe is dedicated to recognition – that is, to making judgements about whether the content of our current experience is or is not familiar. If this memory system becomes active spontaneously, as it might in an epileptic seizure, whatever novel events we are currently experiencing will be mislabelled and experienced as familiar. The eerie quality of *déjà vu* stems from the fact that we are unable to recollect any previous encounter with a set of circumstances that we feel we recognise intimately.

One line of support for this idea comes from work with an unusual group of children whose hippocampi were injured very early in their lives. The hippocampus, at the heart of Broca's limbic lobe, is particularly vulnerable to interruption of its blood or oxygen supply. It can therefore get into difficulties in the course of prolonged or complicated labour that deprives the baby's brain of oxygen for a while. From what we know of the functions of the hippocampus, one might predict that these children should have very poor memories. Faraneh Vargha-Khadem, a highly productive and original developmental neuropsychologist working in London, has studied such children as teenagers, with fascinating results.

Their memories are indeed very poor, in a certain sense. Jon, for example, Faraneh Vargha-Khadem's most closely studied patient, remembers little about his life from day to day – what he was wearing, or whether he ran into a friend yesterday, or if he needs to go to the dentist tomorrow. His recollection of the past, and likewise his prospective 'memory for the future' are poor. But, remarkably, this has had little impact on his general intellectual development and his functioning at school. Jon learns efficiently, but he seems to do so using part of his memory system outside the hippocampus, the part, Vargha-Khadem believes, that in most of us supplies the sense of recognition and familiarity, in the absence of recall. This system allows Jon to learn about the world, acquiring what psychologists call semantic memories, despite being unable to remember the individual encounters through which he accumulates these memories. Studies of patients with epilepsy in whom *déjà vu* occurs as part of the seizure suggest that this 'parahippocampal' brain region, the likely source of Jon's surviving memory, and of our capacity for recognition, is indeed also the source of the experience of *déjà vu*. We are at the limits of our current knowledge here, and the neurological explanation of *déjà vu* is still under debate.

We can be certain, though, that memory is continually at work. Pause for a moment. Look around. Ask yourself how things would seem if you suddenly lost your memory for the events of the last month. You might be completely mystified, like Jed, by your surroundings. At the very least, you would be puzzled – why is that novel lying on the table? Who brought me those flowers? How did I get *here*? We constantly interpret the present in terms of the past. It should not surprise us too much that the mechanisms of memory occasionally misfire, summoning a powerful sense of pre-existence.

CHAPTER 8

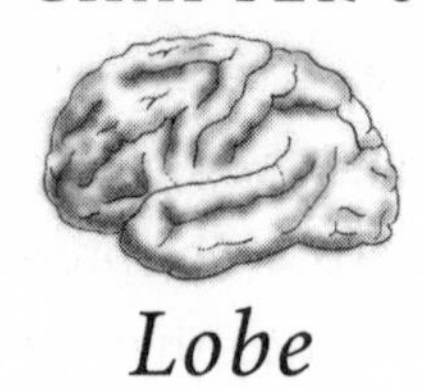

Lobe

The Art of Losing

The art of losing isn't hard to master. . .
Elizabeth Bishop, 'One Art'

REMBRANDT IN LONDON

The brain's networks, like the memory network we have just encountered, thread between its major visible subdivisions, the four lobes of the cerebral hemispheres. Though the boundaries between the lobes are artificial, convenient partitions of a continuous whole, these four great territories correspond to major aspects of the mind. This chapter introduces the lobes of the brain, and their roles in the life of the mind.

If you are ever passing Kenwood House, perched on the edge of London's Hampstead Heath, look in for half an hour. This handsome eighteenth-century mansion contains a small, but wonderful, collection of paintings. Tucked away close to the end of the tour, waiting broodily for your attention, is one of the world's great pictures. In Kenwood's Dining Room, Rembrandt van Rijn, by this stage of his life impoverished and embattled, looks out at his visitors from a self-portrait that expresses an almost unbearable self-knowledge: wary, defiant, meticulously truthful, he was, by then, master of sorrow, of ruin, of mortality: master of loss.

He had a lifetime of painting behind him by 1662 when he painted this self-portrait, one of the last in his lifelong series. Even at the outset, some thirty years earlier, despite his love for gaudy costumes and disguise, his self-portraits somehow anticipated trouble: nothing, the paintings say, nothing will last, nothing endure; youth, good fortune, beauty, all these fade. And so they did. The 1662 self-portrait was painted in the midst of personal disaster – his relationship with his housekeeper, mistress and model, Heindrijke, condemned by the Church, his self-made fortune spent. In that same year Rembrandt was forced to sell his first wife, Saskia's, grave; the following spring, Heindrijke herself died of the plague, and was buried in an unmarked plot. The thousand fragile links that connected Rembrandt with his previously successful life were loosening, terminally. His art was large enough to encompass loss. For others, loss is the unexpected source of new creation. This chapter describes such a case of gain through loss, the case of one of Rembrandt's countrymen, afflicted by a strange, creative, malady. I met Jan late in its course, but his family helped me to piece together the tale of its beginning.

SPRING BLOOMS

Returning from the fields after a walk with his grandson Pieter, some years ago, Jan, now in his early sixties, was uncharacteristically perturbed. The farm had grown the same varieties of tulip for a decade. But when Jan tried to name the blooms for Pieter as they passed them, they wouldn't come to mind. It was as if, somehow, he no longer *knew* the flowers, though he found them more beautiful than ever, vigorous, glossy, bold. The tulips' colours were at their most vivid just before harvesting: deep crimson, delicate lilac, sheer white and black as night. Jan consulted his *Dictionary of Flowers* when they got home: the names were not as familiar as they should have been, and were not on the tip of his tongue. He felt as if he were losing some connection with himself – as if an invisible thread had snapped while his back was turned: something important had slipped away.

Jan's farm was one of the oldest in south Holland. Jan could travel round it in his mind's eye without taking a step from the house. The fields where he walked with Pieter are bordered by small waterways. Lines of willows run alongside them. Jan's land extends for half a mile to left and right. Opposite the tulip fields stands a fruit and vegetable garden, sheltered by a low wall, large enough to supply the household's needs, with some to spare for friends – a cornucopia of potatoes, onions, tomatoes, lettuces, raspberries and currants. To the south, Jan could visualise the lawn where, these days, Pieter played and the family liked to eat outside on balmy evenings.

The bookshelves in his study housed an untidy collection of favourite novels, some poetry, maps, histories, art books. Jan had always admired the painting of his countrymen, especially the masters Bosch, Brueghel, Rubens, Rembrandt and Vermeer, though he was also fond of the shivery winter scenes and gusty sky- and seascapes that fill the nation's galleries. He sketched, and every so often found time to paint a watercolour of some corner of the farm. In a corner of the crowded room stood a harpsichord, a present from Jan's father years before; Jan played well, though only once in a while and mainly for himself. Bach's Preludes and Fugues had been his lifelong companions. Occasionally his daughter, Pieter's mother, now a professional flautist working in Amsterdam, would join him in the evening for duets. Jan sat at the harpsichord while Carolina stood, graceful against the faded tapestry, woven in russets, browns and blues, that hung on the wall above the instrument: these were amongst the fondest moments of their lives.

Jan had known something was wrong for a while. Each time he sat down to write he found that the more difficult words escaped him – and if they came, their spellings, when he checked them in the dictionary, looked outlandish. His trouble finding the tulips' names this morning underlined the gloomy message: something most peculiar was happening to his memory. But his pleasure in the farm, in the tulips on the mantelpiece, in the magnificence of Bach, was undiminished. In fact, Jan was almost sure that it had grown.

LOBE BY LOBE

We are inveterate map-makers. We map our neighbourhoods, from the local to the cosmic, as a matter of course, but map-making metaphors come naturally to our lips in quite different kinds of endeavour: we 'anatomise' a problem, learn our way around a language, 'map the mind'. Geography may not reveal the inner essences we seek – but we are ever hopeful that it will, and a good map undoubtedly makes a helpful start to our exploring. When the seventeenth-century Oxford physician Thomas Willis published his study of the brain's anatomy, *De cerebri anatome*, in 1643, with its monumental illustrations by his astronomer-architect friend Christopher Wren, Willis declared that his intention was to 'unlock the secret places of man's mind'. Such an ambition might seem unexpected in an anatomist, but Willis shared the belief expressed by the French neurologist Paul Broca two hundred years later, that the 'great regions of the brain correspond to the great regions of the mind'. At least to some extent most modern neuroscientists would agree.

Viewed from the side, the surface of the exposed brain is composed mainly of folded cortex, with a lip of the cerebellum protruding to the rear, the brain stem descending beneath where it fuses with the spinal cord. The cortex, from the Latin for 'bark', is a soft, gently pulsating landscape of hills and valleys, 'gyri' and 'sulci' in the anatomist's jargon. Their geography varies somewhat from brain to brain but there is a sufficiency of reliable landmarks to identify a set of major subdivisions, the 'lobes' of the cerebral cortex. These have long been regarded as prime candidates for the material basis of the human mind.

The *frontal* lobe is marked off from the rest of the brain by the central sulcus, a relatively deep cleft running roughly vertically down the convexity of the exposed hemisphere. Immediately ahead of the central gyrus lies the 'motor strip', a strip of cortex containing a map of bodily movements. The space allocated to the parts of our body in this 'motor map' space corresponds to the skill with which we can control them: the hand occupies a larger region than the leg. Immediately behind the central gyrus, a corresponding map, the sensory cortex, processes information about touch and limb position. This sensory map lies in the *parietal* lobe, which divides the frontal lobe ahead from the *occipital* lobe at the rear of the brain. A less conspicuous parieto-occipital sulcus marks the boundary between these. Running forward below the frontal and parietal lobes, separated from them by the sylvian fissure, the fourth of the great cortical territories, the *temporal* lobe lies on the base of the skull.

These regions of cortical 'grey matter', rich with neuronal cell bodies, are massively interconnected, by tracts of 'white matter', great bundles of axons running between them. There is therefore no

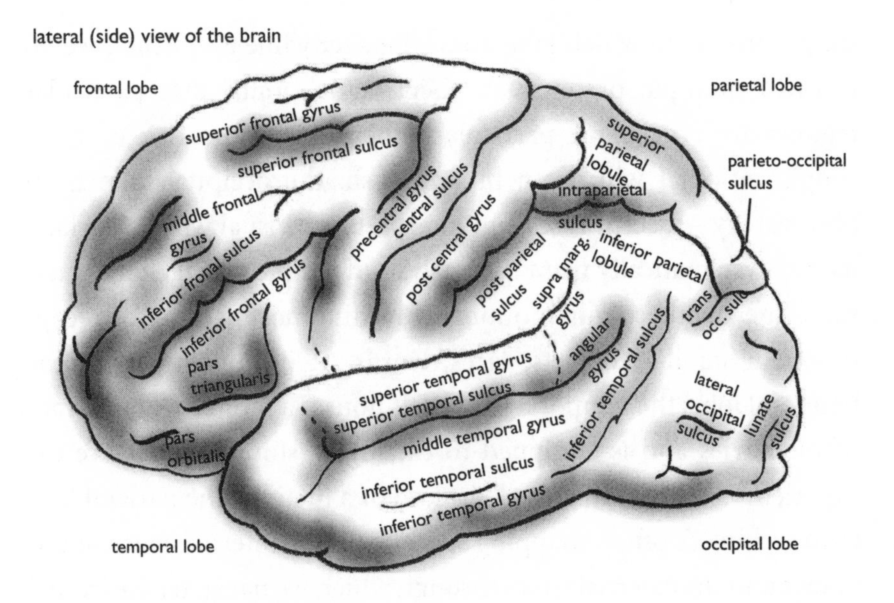

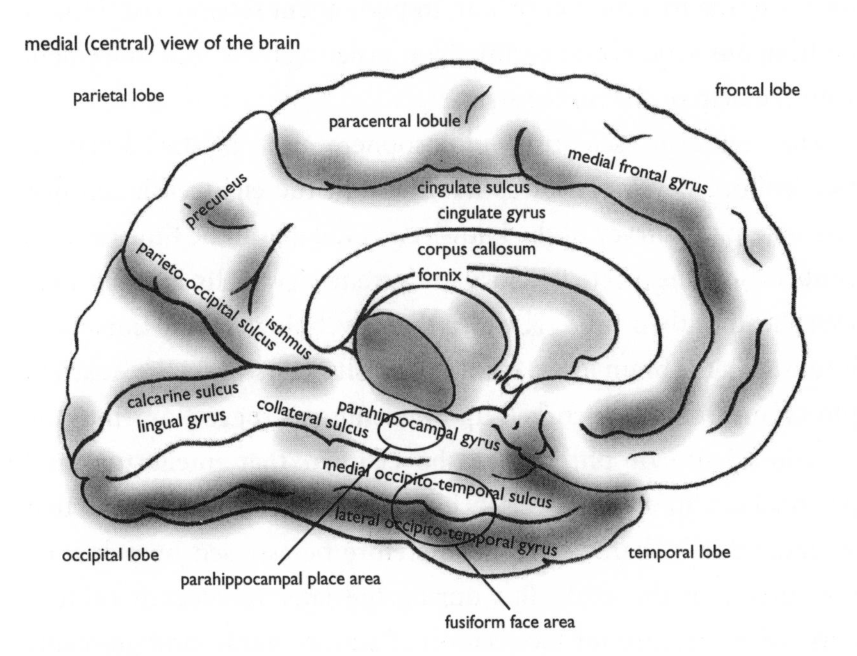

8. The Lobes of the Brain
These views of the left hemisphere show the four lobes of the brain, with the hills (gyri) and valleys (sulci) of the cerebral cortex. The shrinkage of Jan's brain mainly affected the left temporal lobe, shown at lower centre here. The fusiform face area and parahippocampal place area are discussed in the text. The cerebellum and brain stem are not shown here.

simple answer to which lobe does what. Yet while everything we do involves multiple brain areas, there is no doubt that particular regions are crucial for particular functions. At the risk of gross simplification, I will use a broad brush: the occipital lobe is an observatory – it scrutinises and interprets the visual world, working its way systematically from the analysis of simple visual characteristics – line, depth, colour, motion – in its hindmost part, to the recognition of familiar objects, faces, words, at its junction with the temporal lobe; the temporal lobe itself houses a monumental library of memories and is equipped to catalogue, store and retrieve the experiences and accumulated wisdom of a lifetime; the parietal lobe is an architect's office, mapping space, both the internal space of our bodies and the external space through which we navigate; the frontal lobe is home to a cabinet of war, in permanent session, continually plotting our actions and eventually translating these into movement with the help of the motor strip.

The Swedish scientist, philosopher and mystic Emanuel Swedenborg developed ideas like these in the eighteenth century. His systematic survey of the knowledge available at the time reached conclusions unexpectedly close to current views. His insights were well ahead of their time. The idea that psychological functions were localised in the brain lost credibility for fifty years when the Austrian physician Franz Gall worked up Swedenborg's suggestions into the pseudo-science of phrenology: he claimed that intellectual and personal qualities were precisely localised in areas of cortex, and that an individual's character could therefore be assessed by palpating the surface of the skull. But during the later nineteenth century persuasive evidence for localisation of a more subtle kind gradually accumulated: Broca published his observations on the localisation of the capacity for fluent speech in the left frontal lobe in 1861;

Gustav Fritsch and Eduard Hitzig demonstrated the existence of the motor strip in 1870; Carl Wernicke identified an area required for language comprehension in the left temporal lobe in 1874; Korbinian Brodmann produced a detailed map of the cortex distinguishing around fifty distinct areas in the 1900s, identified using staining methods like those of Golgi and Cajal. He anticipated that the fine variations of structure he had identified must correspond to differences of function. He has been proved at least partly right by research over the past century.

Two much-studied areas illustrate the degree of specialisation of function in the cortex. The 'fusiform face area' lies on the undersurface of the brain, just inside the temporal lobe, close to its junction with the occipital. Work using imaging techniques that reveal the activation of the living brain as it performs a given task shows that this area is strongly excited by faces, especially familiar ones. The same area becomes active if one simply *imagines* a face. Correspondingly, damage to the fusiform face areas on both sides of the brain results in an inability to recognise faces, face-blindness or 'prosopagnosia'. Close by lies a quite separate area, the parahippocampal place area. This region is excited by the sight of buildings and distinctive locations. Like the fusiform face area, imagination also excites it, but this time the imagery required is of the Eiffel Tower or Wall of China rather than Marilyn Monroe.

Talk like this is sometimes condemned as contemporary phrenology – and certainly it would be crazy to imagine that these areas *alone* can identify faces or places. Their connections with other brain regions, some in the cortex, some deeper in the brain, are crucial if they are to perform their role. And the discovery that they are tied to these roles falls far short of telling us what kind of neuronal computation they perform, *how* they accomplish their tasks. Nonetheless, these findings show that these areas are key nodes in the neural networks that allow us to recognise faces and places – necessary, if not sufficient. They exemplify the degree of cortical localisation of mental function in current research. The more basic discovery, that the lobes of the brain have distinctive and somewhat specialised roles in our psychology, is absolutely fundamental to contemporary neurology. It also allows us to understand the source of Jan's peculiar troubles.

PURE FORM

Two years had passed. As language gradually deserted Jan, he withdrew increasingly from contact with his friends and even his family, spending his time during the warm spring and summer in the fields, with his paints. Pieter, his grandson, understood that Jan no longer wanted to speak much. They formed a taciturn alliance, Pieter running back to the farmhouse to fetch a jug of water, a spare brush or to explain where they were going. They enjoyed each other's company without the need for speech.

Jan appeared bewildered now in the midst of crowds and conversation. But his puzzlement disappeared when he painted: he worked with a contented concentration, for hours on end, only returning to eat, or when the light failed. He produced a painting more days than not. They pleased the eye, these delicate descriptions of the farm, taking in the tulip fields, the narrow waterways, the windmill, willows and the church. But the paintings that startled his family were in a different vein. Jan had found a box of oils in his study, opened only three or four times before. With these, when it rained, he painted still lifes, often a vase of tulips, set above the fireplace or standing on the harpsichord. Somewhere he found reserves of technique no one, including Jan, had known he possessed, to create a poetry of form and colour – his tulips unfolded glistening leaves and lovely blooms above the harpsichord, against the faded tapestry. Sometimes a tax bill or a pile of euros lay alongside the vase, a laptop winked from the desk; Jan painted these with the same exactitude as the flowers, working in a world of shape and feeling cut free from everyday worries.

Jan's experience – a flowering of artistic talent in the face of a dementia – is unusual but not unique: artistic talents can flourish unexpectedly as a result of brain disease. The American behavioural neurologist Bruce Miller has taken a special interest in cases like Jan's. In 2000 he described twelve patients who acquired or maintained impressive abilities in music, invention or visual art in the course of their dementia. Several had been exceptionally talented in their earlier lives, though often in other areas. Some of his patients won prizes for work produced after the onset of their illness. Most of his patients pursued their new artistic interests to the exclusion of the other normal activities of life. The common thread in their disease was early loss of the ability to communicate through language, together with shrinkage of the left temporal lobe.

Cases like Jan's raise two compelling questions. Artistic creativity – like language – is one of the key capacities that distinguishes us from other animals. If it can come apart from the ability to use language in people with brain disease, what are its underpinnings – what kinds of understanding and skill *does* it require? And what does this divergence of the use of language and the creation of art imply about how experience and behaviour are represented in the brain?

Beyond a doubt, Jan's paintings proved he could perceive his surroundings clearly: they faithfully reflected both the individual elements of the scene – the tulips and the tapestry – and their spatial relationships. More, he could clearly hold them in mind while he conveyed their essence on to canvas, and he had retained the skills required to choose and form the brush strokes this required. In the psychologists' vocabulary, his painting implied that visuo-spatial perception, visual working – or 'short term' – memory and 'praxis', the capacity for skilful action, were all intact.

But Jan's painting implied something more profound: the survival of a more uniquely human faculty – the *detachment* that frees us from the urgency of the here and now, allowing us to stand back, admire, contemplate and, if we are skilled enough, describe what we experience. This same detachment allows us to travel mentally into the past, and to conceive the future, to imagine other worlds and other times. It lies at the heart of art and science. It depends, particularly, on our well-developed frontal lobes, regions that remained intact, so far, in Jan's deteriorating brain. All these abilities lived on in Jan – and something more. 'To be free is often to be lonely' – Jan retained the desire to share experience, one of the most poignant of our human urges, reflecting our need to communicate across the gulf opened up by our uniquely human detachment.

Jan still liked to play Bach with Carolina when she visited. He spoke little now, but his musicianship was undiminished. In fact, it had a more light-hearted, inventive quality than before. One evening, standing a little behind Jan as he sat at the harpsichord, Carolina, to her bemused delight, found herself forced to improvise a flute accompaniment to the 49th Prelude and Fugue – one that Bach himself had never quite found time to put on paper.

A BROKEN SYMMETRY

The root of Jan's troubles, like those of Bruce Miller's patients, lay in the left temporal lobe. His brain scans revealed both local loss of brain substance and a reduction of activity. The foremost part of the lobe, the 'temporal pole', was most affected. Can this explain Jan's combination of language loss and artistic gain? There are reasons to think that it can.

We are symmetrical creatures: most of our organs are paired. In some cases the two members of the pair are twins with nearly identical structure and functions – the two kidneys, or lungs, for example. In others, the organ's two halves have divided their labour. The left side of the heart supplies oxygenated blood to the body, via our arteries, while the right side receives blood lacking oxygen, from our veins, and sends it on to the lungs for refreshment. While both sides of the heart act as pumps, the details of their anatomy reflect their differing tasks. In most of our mammalian relatives, so far as we know, the brain's two hemispheres are mirror images, the left side of the brain controlling the right side of the body, and vice versa. There are hints of division of labour, of incipient specialisation, in the brains of our closest living relatives, the apes. But among our hominid ancestors, some time during the five million years of evolution that separate us from the apes, matters changed radically: the symmetry of brain anatomy and function were broken once and for all.

The topic is hugely controversial. There is no doubt about the bare fact of asymmetry, but much disagreement about how best to define and interpret it. The most indisputable asymmetry relates to language function, particularly in adults. If you are right-handed it is overwhelmingly likely that your left hemisphere largely looks after your use and understanding of language. If, like around 10 per cent of the population, you are left-handed, it remains more likely than not that the left hemisphere is dominant for language, but there is a distinct possibility – around 15 per cent – that the normal asymmetry will have reversed, so that your right hemisphere leads in language, and a similar chance that the hemispheres share this work, a state known as 'mixed dominance'. The dominance of the left hemisphere is reflected in an anatomical asymmetry, with subtle expansion of language-related areas on the left in comparison to their counterparts on the right. And the left hemisphere is dominant for more than language: controlling the right hand, the normally dominant hand, it takes the leading role in praxis, the capacity to perform skilled actions like using tools and writing.

The area of the left temporal lobe that was shrinking in Jan's brain scans is linked to a particular aspect of language – the ability to recognise and name the 'categories' we use constantly, without a conscious thought, through which we conceptualise our world. You are very likely sitting on a *chair*, and near a *table*. Your extensive knowledge of these items, both of their usual properties and of their names, resides substantially in your left temporal lobe. If this is damaged, you accordingly lose this fundamental database of knowledge about language and the world.

What then of the non-dominant, subservient, right hemisphere? Its specialised functions are less clear-cut than those of its left-sided partner. It plays a more important part than the left hemisphere in some visuo-spatial functions, like the recognition of faces. It is biased towards the perception of 'gist' rather than detail, of the whole rather than its parts. There is evidence that it is biased towards knowledge of individuals, for example your mother or your beloved, rather than knowledge of the broad categories, like 'relative', favoured by the left hemisphere. In keeping with its interest in the personal, the right hemisphere is more involved than the left in interpreting emotions and achieving empathy. It does play a role in 'processing' language, but this is consistent with its other tendencies: it allows us to produce and interpret prosody, the lilt of language that conveys the emotional tone of our speech, whether jubilant or downbeat. And, as you may have guessed, it leads by a short head in one more domain of perception: our grasp and enjoyment of music.

The brain's asymmetry *may* be the key to human uniqueness. Certainly language and tool use, two key functions linked to the left hemisphere, are among the defining human skills. The division of labour between the hemispheres may have created novel, fertile, computational possibilities in the brain. Disturbance of the normal partitioning of function between the hemispheres appears to contribute to the underlying disturbance of brain function in schizophrenia – language is less securely anchored in the left hemisphere than normal – and might be its root cause. The hunt is on, therefore, for the genes that bias the hemispheres towards asymmetry of function. We are not the only 'lopsided' creatures: the canary's song is controlled from within a single hemisphere, suggesting that skilled vocal communication may benefit from a single control centre in other species too. And our tendency to use our left hemisphere for language can be gainsaid: after damage to the left hemisphere early in life, the right hemisphere is capable of taking on the task. But not so for Jan: during his childhood his left hemisphere assumed its normal role, providing the archive of knowledge that enabled him to classify his world and name its contents. The erosion of this hemisphere, in his late middle age, gradually robbed him of that knowledge – but in his case, at least for a while, this process was simultaneously releasing other skills.

RELEASE

Oliver Sacks has written eloquently of the manifestation of neurological disorder in *excess*, in the ebullient release of function. Too exclusive an emphasis on the losses inflicted by disease, Sacks points out, can suggest that the brain is a mere 'system of capacities and connections', a passive machine, reactive at best, defective in disease. Studying the more exuberant phenomena of disease helps to reveal the 'life of the mind' in all its productive – and over-productive – potential.

There is no shortage of examples. Sacks describes the impetuous punning, twitching and cursing of Tourette's syndrome in 'Witty Ticcy Ray'; the contented 'release of thought and impulse' in Natasha K, a woman of 90 with a syphilitic infection of the brain; the constantly shifting inventions – 'confabulations' – of patients with the alcohol-induced disorder of memory, Korsakoff's syndrome. Sometimes the release of function can be pinned to a single event, like the visual hallucinations of Charles Bonnet syndrome which follow the onset of blindness, the ever-productive visual brain creating visual experiences spontaneously when it no longer has the opportunity to create them in league with the eye.

Release occurs at the personal level too. The earnest advice that the problems we encounter in our lives stem, not from our circumstances, but from ourselves, is usually well meant and sometimes right – but it can be completely wrong. The difficult girl who blossoms when her siblings leave home, the child who flourishes when she changes school, prove that our behaviour is the outcome of a dense web of influences: sometimes what's needed is another chance, a change, some space in the light to grow.

The flowering of Jan's art and music in the last few years of his life provides a case in point. Through most of his adulthood, Jan's equable personality had equipped him admirably to manage the farm, care for his family and keep an eye on his elderly parents. He allowed himself the occasional hour of relaxation with his sketch-book, or the harpsichord, but these were brief holidays from the everyday business of life. In retrospect, his articulate, civilised persona must have oppressed the shy potential that dwelt within the right side of his brain: a dormant playfulness, a creativity and thirst for beauty were constrained by the more abstract, businesslike, qualities of his dominant hemisphere. As Jan's left temporal lobe withered, the checks and balances that had made him such a constant citizen gradually fell away: after sixty years of benign inhibition, his right hemisphere was erupting into life.

THE LOSS OF MIND

It's evident
the art of losing's not too hard to master
though it may look like (*Write* it!) like disaster.
Elizabeth Bishop, 'One Art'

The eruption and the flowering were all too brief. They lasted as long as the changes in Jan's brain were confined to his left temporal lobe. But within a couple of years the process had spread – into the frontal lobes and across to the right hemisphere. As it did so, the shifting balance of power in his brain began to work decisively against him: he lost the drive and vision that had so powerfully fuelled his work. Like Rembrandt, though for different reasons, Jan suffered a terminal loss of the connections that had allowed him, for a while, to rescue part of his experience through art.

Jan suffered from fronto-temporal dementia – FTD – an uncommon form of dementia overall, though it accounts for about a quarter of cases starting below the age of 60. Its manifestations depend on which lobe and which side of the brain are first affected: the 'frontal' type presents with particularly marked changes in personality and behaviour – failure to respect social niceties, loss of interest in dress and hygiene, decline in sympathetic concern for family and friends, irritability or withdrawal. The waning of language and conceptual purchase in Jan's case is typical of the form of FTD that starts in the left temporal lobe, christened 'semantic dementia' by Julie Snowden, a British psychologist, and by my mentor, the behavioural neurologist John Hodges. This is the variety most commonly – though by no means always – associated with a surge of creativity like Jan's. The right temporal lobe is affected less often: in such cases there can be an initial loss of knowledge about people: first difficulty in recognising their faces, gradually followed by loss of the underlying knowledge of who familiar people are. We do not know why the frontal and temporal lobes are picked out by this disorder, or why it is so often asymmetric, apparently attracted, in particular, to the left temporal lobe. Several kinds of change can be seen in the brain beneath the microscope in FTD: all have in common the loss of neurons from the frontal and temporal lobes.

'Dementia' is a vague diagnostic term, implying the simultaneous decline of several intellectual abilities – for example memory and language and problem-solving – with an adverse impact on everyday functioning. Its causes are legion, but the most common type, by far, is Alzheimer's disease. This also has its distinctive features. Its first symptom is usually difficulty with new learning – and hence in recalling recent events. The underlying process is a patchy accumulation of deposits or 'plaques' of a protein called 'beta amyloid' in spaces between the brain's cells, together with the appearance of 'tangles' of 'tau' protein within them. It begins, not in the six-layered neocortex of the temporal or frontal lobes affected in fronto-temporal dementia, but in the relatively primitive one- to three-layered cortex of the medial temporal lobes. The process gradually spreads into adjacent regions of the brain so that, over time, language and established knowledge are eroded, as well as the first-affected ability to form new memories.

Alzheimer's disease quite often coexists with changes in the brain due to wear and tear in the arteries that provide its blood supply. Sometimes these 'vascular' changes can cause dementia independently. This process, occurring alone or together with Alzheimer's disease, is the second most common current cause of dementia. The third most common type is more distinctive. In 'Lewy body dementia' the ability to concentrate varies markedly from hour to hour and day to day, with alternating periods of confusion and lucidity. There may be vivid visual hallucinations, often of human figures and animals in and around the home. There is often some sign – a tremor, or slowness or stiffness – to suggest the beginnings of Parkinson's disease: 'Lewy bodies' were first identified as the hallmark of Parkinson's disease: these are rounded accumulations of protein within the dopamine-producing neurons of a region of the brain stem called the 'substantia nigra'. The discovery that these bodies could spread around the brain, causing this distinctive type of dementia, sometimes with little sign of Parkinson's disease, is recent.

Above the age of 65, one in twenty of us suffers from incipient or established dementia; over 80, the figure approaches one in three. Given our lengthening life spans, and the growing proportion of elderly among us, the disease threatens to become rampant. The ghastly 'dementors' who chill the blood of Harry Potter, the wraith-like 'spectres', who feed on human consciousness in Philip Pullman's *Northern Lights*, are scarcely more alarming. Yet dementia at close quarters, like most illness, has a human face. Jan died not long after he lost his enthusiasm for painting. Just a few months before, he played duets for the last time with his daughter. Rembrandt would surely have understood, and grieved: nothing endures, not youth, nor wealth, nor sumptuous intellect.

CHAPTER 9

Psyche

Betrayal

And that this place may thoroughly be thought
True Paradise, I have the serpent brought.
Donne, 'Twicknam Garden'

THE SUBTERFUGE OF IGNORANCE

We have worked our way through the tiers of the brain's organisation, ascending from invisible atom to palpable flesh. We have moved to and fro between several kinds of description that illuminate each other – descriptions of structures within the brain, explorations of their workings, evocations of the experience and behaviour they make possible. All this lies comfortably within the territory normally occupied these days by neuroscience. It provides the scientific basis for clinical neurology, which deals with the brain's disorders. But neurologists routinely encounter a rather different kind of problem, with which we are less comfortable. Put simply, the problem, a common one, is this: about one third of the patients who come to see us with symptoms that suggest a neurological disease turn out, on careful assessment, not to have one at all. Instead, there is often – though not always – evidence of some psychological predicament lurking in the wings.

Neurologists – and physicians generally – are not well trained to deal with this situation. We tend to feel uncertain with such patients and their problems because we lose our professional bearings. The tools we gain assiduously during our apprenticeship, our skills in neurological assessment, are invaluable in diagnosing the neurologically defined dysfunctions of the brain; but much less help in fathoming the problems of the psyche. Common sense and fellow feeling can come to our aid – as can more formal training in psychiatry, but neurologists seldom have this.

On a first encounter it is often difficult to be sure whether a problem is primarily neurological or psychological. Alison's case, at the start of this book, illustrates the problem: the cause of her chronic fatigue turned out to be decidedly neurological, to my embarrassment and surprise. Neurologists therefore rightly hesitate to assume that there are 'psychological' causes for disorders they simply don't recognise or understand: we are cautious of using the term, in the words of the seventeenth-century physician Thomas Willis, as the 'subterfuge of ignorance'. But the converse risk – of mistaking a primarily psychological problem for a neurological one – is just as serious, potentially leading to expensive and unnecessary tests, to dangerous treatments and even, occasionally, to the lawcourts.

The famous Canadian neurosurgeon Wilder Penfield described neurology as 'the study of mankind itself'. If so, neurologists should not expect to ignore its private life. The case I am about to describe reminds us that the medical and the personal are often intermingled. Doctors do better not to switch off their everyday emotional antennae when they slip on their white coats.

COLLAPSE

I first met Jenny towards the end of a long clinic. The National Health Service allows us fifteen minutes for a follow-up appointment, a more generous allocation than the family doctor's, but still brief when the task is to explain a diagnosis of multiple sclerosis or motor neuron disease – or to try to get to know someone whose difficulties may turn out to be rather private.

I scanned Jenny's notes before I called her. Her file was a little thicker than usual. I soon realised that although I was 'following her up' we had, in fact, not met before. She had been in hospital a couple of months earlier, on the infectious disease ward, a junior doctor on the team had reviewed her, and I was picking up the threads. The discharge summary was helpful – but a little puzzling. She had been admitted with a bad headache. Meningitis was suspected, but the relevant tests – a scan and a lumbar puncture to examine the spinal fluid – had given normal results. Migraine seemed the likeliest explanation but she had been treated with antibiotics, all the same. She had been slow to pick up. A week after admission, her left arm and leg became weak, and her gait unsteady. Two weeks into her illness, her mother had been admitted to hospital with a heart attack and Jenny had hobbled back to her family, to help look after her kids – and help her dad. And now here she was; better, I hoped.

My first impression, to be honest, was that Jenny was extremely pretty – petite, pale, with blue-green eyes, dark hair swept back and girlishly collected in a ponytail, a scattering of freckles round the eyes, a neat, attractive, figure. But she wasn't happy. She shook my hand without enthusiasm, and didn't return my smile. I asked her how she was. Her replies to my questions were guarded – but so what? She didn't know me, and she'd been through a lot recently. She still wasn't well: her headaches had continued after returning home. She had blacked out three or four times since leaving hospital, bruising her arm badly. Her left side felt numb and clumsy. Ah, dear: this was definitely more than could really be dealt with in a quarter of an hour. But my impression was that the elements of this illness just didn't add up – there were the makings of too many diagnoses: the possibilities of meningitis, migraine, epilepsy and multiple sclerosis had all been considered, but she clearly didn't have all of these and it was doubtful that she had any.

A brief examination of Jenny's arms and legs – muscle tone, strength, reflexes – didn't reveal anything much. She was coping at home, just about. I promised to look more carefully through her notes, review her scan, and see her again within a few weeks, before planning the next step. I had a hunch that there might be something in the background that Jenny was keeping to herself.

A couple of minutes after she left my room, I was called out by the nurse in charge. Jenny was lying face down between the automatic doors of our Outpatient Department. She was pale and still. Her eyes were closed. She had a pulse and was breathing. We pulled her unconscious body gently into the waiting room – and waited. One of the nurses checked her blood sugar with a finger prick. It was normal. Jenny began to move a little and to groan after a few minutes, and we helped her to a couch. Another patient's folder had appeared outside my room and I went off to finish my clinic. By the time I was done Jenny had taken herself off home. What could be wrong with her?

THE UNQUIET WOMB

I conjure thee, O womb, by our Lord Jesus Christ, who walked over the sea with dry feet, who cured the sick, who expelled the demons, who brought back the dead to life . . . by whose plight we were healed, by Him, I conjure thee not to harm that maid of God, . . . not to occupy her head, throat, neck, chest, ears, teeth, eyes, nostrils, shoulder blades, arms, hands, heart, stomach, spleen, kidneys, back, sides, joints, navel, intestines, bladder, thighs, shins, heels, nails, but to lie down quietly in the place God chose for you, so that this maid of God . . . be restored to health.

Tenth-century prayer

Hystera, the Greek for the uterus, the womb, has lent its name to a disorder, 'hysteria', that has been recognised in every place and period, a disorder with a baffling multiplicity of forms, challenging understanding of the relationship between body and mind, lying in wait for the unwary physician, who is equally at fault if he fails to recognise its presence or attributes it in error. Its symptoms, as classical authors recognised, are immensely varied: they include loss of speech, paralysis of the limbs, fatigue, loss of sensation, mysterious pains, violent convulsions, bland loss of consciousness; indeed it can manifest itself in just about any disturbance of experience and behaviour. Symptoms like these are the bread and butter of neurology, but in patients with hysteria they are not a straightforward outcome of neurological disease.

The Egyptians, and later the Greeks, believed that the confusing manifestations of 'hysteria' resulted from the wandering of a restless womb around the body: rising to the heart, it would incite panic, to the throat, a sensation of choking, to the brain, a sense of lethargy. Their remedies were correspondingly designed to drive it back from the parts of the body to which it had wandered, repelling it with unappetising substances like excrement and tar, enticing it home with delicate fragrances – myrrh, pine dust, frankincense.

In the Middle Ages, symptoms that might, in the ancient world, have been attributed to hysteria came to be regarded as a sign of witchery, possession, 'unholy copulation', requiring not medical treatment but exorcism, the pillory or the stake. During the seventeenth century, as modern forms of scientific thinking began to emerge, these symptoms were gradually reclaimed by medicine. In 1603 a physician, Edward Jorden, gave evidence in favour of the fourteen-year-old Mary Glover whose apparent fits, intermittent blindness and paralysis had all been attributed to witchcraft. His treatise *A Briefe Discourse of a Disease Called the Suffocation of the Mother* – 'Mother' referred to the womb – argued that her symptoms, and others like them, resulted from natural causes rather than supernatural or magical disease. But the link to the womb was growing more doubtful, not least because of the observation that men, as well as women, are subject to hysteria. Medical opinion gradually relocated the source of hysteria to the brain. Charles Lepois, a seventeenth-century French physician, wrote: 'All these symptoms come from the head.'

In the course of the nineteenth century this view – that hysteria was the result of a primarily physical disorder located in the brain – was reformulated once again as physician-psychiatrists, like the father of French psychiatry, Philippe Pinel, focused their attention on the mind. The move from physical to psychological explanations for hysteria culminated in Sigmund Freud's excursions into the unconscious, especially the unconscious depths of sexual feeling. At the outset of his work, Freud was himself unaware of the long lineage of such ideas, symbolised by the association between hysteria and the womb. Plato had written, in the *Timaeus*, more than two thousand years before: 'The womb is an animal which longs to generate children. When it remains barren too long after puberty, it is distressed and sorely disturbed, and straying about the body and cutting off the passages of the breath, it impedes respiration and brings the sufferer into the extremest anguish and provokes all manner of diseases besides.'

A SECOND LOOK

There are times when we must not look back. Sometimes a backward glance is just too painful. Sometimes, as for Orpheus, who lost Eurydice for ever when he turned to catch a glimpse of her following form, the glance has been forbidden by the gods. But doctors, generally, are well advised to turn back and look again, and then again.

I had arranged to see Jenny at the end of a clinic, to give us a little more time. Her headaches, the numbness and tingling in her left arm and leg, her blackouts, all continued. As a rough but useful rule of thumb, the chances of finding clear-cut physical disease diminish as the number of symptoms increases. Jenny's symptom count was on the high side. Applying this rule, I wondered, again, whether there might be something in the background that she was keeping to herself, some key that could unlock the secret of her illness. I probed gently for a point of entry to her story. I asked some questions that are always relevant to illness, because illness is always rooted in psychological soil – questions about her appetite, sleep, energy levels, her concentration and memory; if she could take pleasure in life, whether she was anxious. She answered all these questions readily enough, but the answers took us nowhere. No, she was not enjoying herself much – but who would, in her circumstances?

I examined her more carefully than before, still hunting for a clue. Jenny limped over to the couch. She seemed to be weak in the left arm and leg, but the weakness wasn't consistent, and at times her limbs collapsed abruptly under pressure. Inconsistency of strength and 'collapsing weakness' generally point away from neurological disease. There was another sign suggesting that Jenny's problem was more expressive of distress than of disease. When asked to press her left leg into the couch, she seemed weak – I lifted her leg easily. But when asked to raise her right leg, the normal one, into the air, the hand I had left tucked in beneath her left leg couldn't raise it an inch. This reflects the normal, automatic, coordination of movement in our legs: when we bend one leg at the hip, the other automatically straightens (this is easily demonstrated with the help of a friend, or even by examining yourself). It seemed that Jenny's left leg was perfectly strong. Her problem was not one of weakness – but, somehow, one of will.

Jenny and I met up in the clinic on a couple more occasions over the next few months. Her tests were all normal. Her multiple symptoms and the inconsistencies when I examined her were unchanged. Each time we met, I felt sure I was missing the point. Each time my open-ended questions met a wall.

THE IDEA OF PARALYSIS

Some of the most serious disorders of the nervous system, such as paralysis, spasm, pain, and otherwise altered sensations, may depend upon a morbid condition of emotion, of idea and emotion, or of idea alone.

Russell Reynolds, *British Medical Journal*, 1869

Hysteria has been a diagnostic pitfall over the centuries in part because it is such a great imitator of other disorders. Galen of Pergamon, a physician of the second century AD whose writings were immensely influential for over a millennium, recognised this clearly: hysteria 'is just one name; varied and innumerable, however, are the forms which it encompasses'. Over a thousand years later the seventeenth-century English physician Thomas Sydenham echoed Galen's words: 'The frequency of hysteria is no less remarkable than the multiformity of the shapes which it puts on. Few of the maladies of miserable mortality are not imitated by it.' Robert Whytt, in the following century, agreed with the 'sagacious Sydenham': 'the shapes of *proteus*, or the colours of the *chameleon*, are not more numerous and inconstant than the varieties of the hypochondriac and hysteric disease'.

What common thread unites such bewildering diversity? The idea that hysteria might be, in some sense, a disorder of the imagination is one recurring answer. Paracelsus, the itinerant sixteenth-century Renaissance alchemist and physician mooted this idea, writing of a dance-like disorder of movement in children: 'the cause of the disease is . . . an idea assumed by the imagination . . . an imagined idea, based not on thinking but on perceiving'. Remarkably,

Paracelsus even drew attention to the possibility that imagination can work unconsciously: 'In children . . . [t]heir sight and hearing are so strong that unconsciously they have fantasies about what they have seen or heard. And in such fantasies their reason is taken and perverted into the shape imagined.' Discussing an entirely different manifestation of hysteria many centuries later, hysterical paralysis, the physician Russell Reynolds described it as a disorder 'dependent on idea'.

Why should an idea, or the workings of the imagination, impose itself on someone's physical well-being, depriving him of movement or compelling him to move abnormally? Most, like the seventeenth-century Italian physician Giorgio Baglivi, have looked for an explanation in emotion: 'all Men have their own Care, and every one lies under a bitter Necessity of spending almost all periods of his Life in attending the doubtful Events of his Labour. Now this being true, 'tis equally a Truth obvious to all Men, that a great part of Diseases either take their Rise from, or are fed by that Weight of Care that hangs upon every one's shoulders.' Robert Whytt, the Scottish physician, one hundred years later invoked 'doleful or moving stories, horrible or unexpected sights, great grief, anger, terror, or other passions' as the causes of 'hysteric fits, either of the convulsive or fainting kind'.

An illness governed by imagination, provoked by emotion and prone to take on a great variety of forms is not a straightforward 'disease'. Indeed, it is tempting to think of hysteria less as a disease than as a form of behaviour – 'a romance of morals and manners', in the words of Pierre Janet, one of its great French nineteenth-century students. This helps to explain the widely fluctuating attitudes towards its sufferers over the centuries, ranging from humane sympathy to uncompromising blame. I am drawn to the views of Weir Mitchell, an American neurologist and scientist who encoun-

9. Patients at La Salpêtrière
These sketches are taken from the famous series of drawings by Paul Richer of Charcot's patients in the Salpetriere Hospital in Paris in the 1880's. Pierre Janet, mentioned in the text, was one of Charcot's students.

tered hysteria in both men and women during the American Civil War: 'The largest knowledge finds the largest excuses .... The elements out of which these disorders arise are deeply human, and exist in all of us, in varying amounts.'

THE CALL OF THE NIGHT

Several months later, Jenny met with a psychiatrist colleague of mine, a young woman, and told her the story that I had so lamentably failed to unearth. Was it my colleague's sex or her age, the name of the clinic, the passage of time or simply the length of the appointment that allowed Jenny to tell her tale? Maybe Jenny herself partially recognised that her problems flowed from her emotions, and decided that someone like me, who doctors the body, could not be expected to understand her particular form of affliction. What had happened took a while to tell, but it went something like this.

Jenny had married Sean ten years before. They had been childhood friends. Sean had always had a soft spot for Jenny and she had found his company comforting at a difficult time in her life. But there was a problem in the wings that Jenny hadn't ever really squared up to – either with others or with herself.

Jenny was quick to conceive. Her boys, Duncan and James, were already nine and seven. But Jenny and Sean, never the closest of companions, gradually drifted apart, especially once Sean ran up against troubles of his own at work, and started drinking. Beer and the TV soaked up most of his attention in the evenings.

One December night Jenny, feeling dejected, had taken herself for a walk. The boys were in bed, Sean half asleep in front of the TV. She intended to come back within a few minutes, but something had kept her heading into town. Surprising herself, she had gone to a bar, ordered a drink. Sitting, cradling her glass, she looked up to meet another pair of eyes. Susan came across to join her. She was tall, blond, self-confident and she met Jenny's gaze with a wry, amused expression. Jenny found her instantly attractive. Sue's company warmed and excited her: her discontented mood rapidly ebbed away. Jenny had been drowning, and Sue had pulled her back to land.

Within a few weeks, Jenny was spending time at Sue's flat whenever she could steal an hour or two away from her work and family. She had never met anyone before, of either sex, who made her feel as Sue did, so raw with desire and longing. But it wasn't really a surprise that things should be this way. This had been the problem in the wings: Jenny had always been drawn sexually to women much more strongly than to any of her boyfriends, Sean included. Her family had made it all too clear to her as a teenager what they thought of homosexuals. Gays were not welcome at home. Jenny had driven her feelings for women somewhere into the back of her mind – where they had bided their time.

One night she stayed much longer than she'd planned. On the way home, wondering whether Sean would be waiting up for her, and what she could possibly say to him if he was, her head began to throb. A few minutes later, unable to help herself, she vomited into the gutter. She was a mess by the time she got home. Sean woke to hear her retching in the bathroom, curled on the floor, holding her head in her hands. She couldn't tell him what was wrong. He called an ambulance.

A SNARE AND A DELUSION?

The diagnosis of hysteria is controversial in principle and can be difficult to make in practice. It is easy to mistake 'the great imitator' for the disorders that it imitates. The risk is so great that some doctors have cautioned against making the diagnosis of hysteria in *any* circumstances: a British psychiatrist, Eliot Slater, waged an especially vigorous campaign against hysteria fifty years ago, arguing famously that it is 'not only a delusion but also a snare'. He claimed that the outlook for sufferers from hysteria was dire: not because of any intrinsic danger in the – he believed non-existent – condition, but because of the risks posed by the disorders that should have been diagnosed in its place, a mixture of grave neurological and psychiatric diseases.

Slater was right to caution his colleagues against over-casual diagnoses of hysteria. Doctors have a recurring tendency to assume that symptoms they can't make sense of must be 'psychological' – and therefore in some sense their patients' fault. This exonerates us from the need to think, or investigate, further. I have already quoted the words of the Oxford physician and anatomist Thomas Willis, who had noticed this tendency in the seventeenth century: the 'hysterical passion . . . bears the fault of many other Distempers: for when at any time a sickness happens in a Woman's Body, of an unusual manner . . . so that its causes lie hid . . . we declare it to be something hysterical, which oftentimes is only the subterfuge of ignorance'.

But Slater was wrong to argue that the diagnosis should never be made; for by whatever name we call this illness – 'hysterical', 'psychogenic', 'psychosomatic', 'functional', 'somatoform', 'medically unexplained' – it is common, and can, with reasonable care, be diagnosed reliably. Indeed, failure to recognise the presence of hysteria is much more common nowadays than erroneous diagnosis. Given the potency and potential side-effects of modern medicine, the results of such failure, if it leads to vigorous treatment for misdiagnosed disorders, can be distinctly bad for your health.

BETRAYAL BY THE BODY

For seeing we are not maisters in our owne affections,
we are like battered Citties without walles, or shippes tossed in the sea,
exposed to all matter of assaults and daungers, even to the
overthrow of our owne bodies.

Edward Jorden, *A Briefe Discourse of a Disease Called the Suffocation of the Mother*

Jenny was admitted to the infectious diseases ward which was always busy over Christmas, its brave attempt at a festive air belied by the paraphernalia of medicine – drip stands, syringe pumps, cardiac monitors, the resuscitation trolley. The experience was pretty scary. The doctor who eventually managed to see her in the early hours was worried that she might have meningitis. He organised a brain scan, as an emergency, followed by a lumbar puncture, which required three painful attempts to slip the needle into her spinal canal.

Jenny woke the following morning to a grey December dawn. Her head still ached and her back was bruised from the puncture. The consultant came round, reassured her that the results showed nothing untoward, suggested she might have had migraine and advised her to stay overnight as she still wasn't well. When he had gone she tried to ring Susan.

Susan's numbers weren't on her mobile but she'd learned them by heart. It was Sunday, so she called her at home. A soft, female, voice answered. It wasn't Susan's. 'Hi,' the voice said. Jenny was lost for a moment, swallowed, then asked, 'Can I speak to Susan, please? Have I got the right number?' 'Yeah, you have – it's Susan's phone.' 'Is she there?' 'She's gone out to buy the paper.' Jenny paused. 'Who am I speaking to?' 'It's Kirsty – Susan's partner. Can I give her a message?' Jenny felt as if she were hearing this from outer space. 'It's OK,' she replied. 'It's OK. I'll call some other time.' As the phone cut off, Jenny closed her eyes and stared at a kind of double desolation.

Sean had to take time off work to look after the kids, which he couldn't afford. When he brought the boys in to see their mum she was too tired and headachy to cope, so they didn't stay for long. Four days into her admission Jenny had begun to puzzle the infectious diseases team. Besides the worsening of her headache, the light now hurt her eyes. She asked for the curtains to be drawn. While the nurse was helping her to the shower that morning, she had tripped. When she was helped up, her left leg felt weak and numb, and her left hand tingled. She had limped back to bed. The following day the news had arrived that Jenny's mother was in hospital with a heart attack. Jenny stumbled out of the ward that afternoon, using a borrowed stick and wearing a pair of dark glasses to shade her sore eyes from the light.

BODY AND MIND

The relationship between the mind and the body is perennially puzzling. How does your mental life – your thoughts, decisions and emotions – influence the machinery of your brain? Conversely, how do events in your body – like the stimulation of your eye by this page – communicate with your mind? Hysteria, in which a disordered state of mind gives rise to a wealth of physical symptoms, poses just the same problem in miniature. Somehow it has to be true that Jenny's thoughts and emotions caused her symptoms. But how?

Some writers on hysteria have simply accepted defeat. Baglivi, who believed that strong emotions were its usual cause, accepted that the 'Mechanick Way' by which the passions produce disease was beyond human comprehension, 'considering that the most tow'ring Genius's of all Ages, have fatigu'd themselves in vain' upon the problem. He was dissatisfied, presumably, by the suggestions made to date – from the wandering of the uterus to the effects of its pervasive 'vapours', or the irregularities of the brain's 'animal spirits' later invoked by Sydenham and Willis.

Times have changed. We really do understand a good deal now about the looping connections between thought, emotion, brain and behaviour. Imagine . . . you arrive home to find the front door ajar, the hall strewn with clothes from your bedroom, your hi-fi gone: you have been burgled, and your home has been turned over. Your brain's skilful appraisal of the sight that greets you allows you to reach some rapid conclusions. Within less than a second, well before you notice it, your body will be joining in the piece. Your pulse and blood pressure rise as the 'autonomic' nerves that run from the brain to your innards signal the stress of the discovery, preparing you for fight or flight, in case these are required; simultaneously, at the brain's prompting, stress-related hormones – adrenaline, cortisol – are released from the adrenal glands into the bloodstream, further fuelling the increase in your pulse, mobilising glucose for immediate consumption, directing blood flow away from the skin and the gut to muscle, and acting back on the brain to heighten your arousal. Much of this physiological activity enters your experience, shaping what it feels like to be there – whether the gearing-up of the stress response gives you a sense of calm control under fire, or sends you skittering towards panic. The hormones reaching to your brain will also help to shape your memory of events, capturing a 'flashbulb memory', or, if stress is unremitting, shrinking the temporal lobe structure involved in memory formation, allowing you to forget.

The pulsing of the heart, surges of hormone release, the workings of our thoughtful brains, our behavioural responses, all contribute to emotion, and each of these feeds back upon the others. Mind and body ceaselessly interact. So far, so good – but does this help us understand Jenny? How, exactly, did her mixed emotions – of guilt, desire, betrayal – lead to her weakness and blackouts? What was happening in her brain? Was she imagining? Faking? If not – what?

WHAT IS HYSTERIA?

Hysteria is a controversial topic but one thing is for sure: it affects the functioning of organisms – us. We are, at all times, a curious fusion of body and mind, and so it is helpful, for any disorder, to examine its origins and its effects along each of the three dimensions of our being: as bodies, minds and social animals. This is the basis for a 'biopsychosocial' approach to illness – an approach most often adopted by psychiatrists, but applicable to illnesses across the board. We know a fair amount, though far from everything, about hysteria along each of these dimensions.

We understand least, I have to confess, about its biological basis in the brain, the question raised at the end of the last section. Clearly *something* distinctive must be happening in the brain of someone, like Jenny, who collapses inexplicably or fails to move a limb when asked to do so. Over the past ten years several research teams have tried to define the features of brain activity in patients with hysteria. The first report, by John Marshall and Peter Halligan based in Oxford, gave a strikingly clear answer. A study of a patient with recurring, unexplained weakness of her left arm and leg used PET scanning to show that her brain prepared for the movement normally – but, just at the moment that a movement was requested, when she tried to move her paralysed leg, areas towards the front of her brain, linked to emotion, came into play unexpectedly. Could these be the source of the 'unconscious inhibition' of movement that Freud might have held responsible in such a case? Well, possibly, but subsequent work has complicated this appealing picture. At the time of writing it is unclear whether the underlying problem in hysterical paralysis is, so to speak, overuse of our psychological brakes, as the Oxford study suggested, or failure to put the foot down on the accelerator, as other work has concluded. This may well vary from case to case, but state-of-the-art brain imaging techniques now make it possible to approach hysteria experimentally and to open up some intriguing related questions: what happens in the brain of someone who deliberately *feigns* weakness? Or of someone with paralysis induced under hypnosis? And how do these cases differ from hysteria, if they do?

I shall have to keep you waiting for the answers – for they remain uncertain.

We know more about the psychological and social conditions for hysteria. Sufferers report higher rates of adverse childhood experiences, especially sexual abuse, than otherwise comparable people without hysteria. Hysterical episodes are often triggered by injury of some kind, especially injuries that cause panic at the time. Depression and anxiety are commonly present. There is also some evidence for a more specific kind of predisposition, which can manifest itself in several ways.

Sufferers from hysteria are more liable than others to consult their doctors with a wide range of unexplained physical symptoms, like irritable bowel syndrome, or fibromyalgia; they are more likely to have undergone surgery like hysterectomy and appendicectomy, sometimes unnecessarily; they are generally less willing to consider psychological explanations for their symptoms. Some people, in other words, seem to be unusually prone to experience or explain life's difficulties in bodily terms. This tendency can be given a Freudian interpretation: Freud suggested that emotional conflict could be translated into a physical symptom to avoid conscious confrontation with the emotions concerned, the source of the term 'conversion disorder' that is still used to describe problems like Jenny's. It can also be related to the concept of 'dissociation', the notion underpinning the leading psychological theory of hysteria. The idea of dissociation is that in response to emotional stress, the integration of experience sometimes fails, so that we lose contact with part of our inner self, as Jenny, arguably, lost touch with her ability to move her limbs or remain conscious. Dissociation, so the theory goes, can be a helpful learned response to overwhelming stress, for example the terrible stress of sexual abuse by a parent, allowing the victim to pretend, in a sense, that she is no longer there.

AND IN THE END?

What became of Jenny? She and Sean parted. A change of marital status – in any direction – predicted a good outcome from hysteria in one study. Jenny's experience bore this out. She stayed in touch with Susan, now and then, still found her exuberance appealing, but kept her distance, as best she could. She could not entirely trust her. But nor, she realised now, could she entirely trust herself. The past few months had been a crash course in betrayal – her own betrayal of her husband, what still felt to her, fairly or unfairly, like Susan's betrayal of Jenny, and her mind's betrayal by her body. And now? She was coping with the kids, just about, and back at work. I saw her once more after she had told her story to my colleague. She was much better. What did she think had happened, to make her so unwell? Well – Jenny grinned, the first time I had seen her smile – how should she know? I was the doctor.

Does a biopsychosocial explanation work for Jenny's case? Her marriage was loveless and stressed, setting the scene. Her emotions were highly charged. Her encounter with Susan raised the emotional temperature further, throwing her into a conflict between duty and desire. She was guilt-prone, and her feelings for Susan reawakened painful memories of her family's attitude when she was younger. At the time she came into hospital, her body had been 'overthrown by her affections' – though the damage may just have been a migraine. Once she had been admitted, her doctors inadvertently contributed to raising her anxiety still further, seeding in her mind the possibility of serious illness. Then her conversation with Kirsty and her mother's illness, both threatening loss, fell on her at a vulnerable time. Her brain and her body responded, in a way that many people can probably sympathise with, by shutting down some of their normal services. Was Jenny simply pretending to be ill? I doubt it – but she was clearly 'acting ill'.

If hysteria is a kind of performance, is it 'all in the mind' after all? Well: yes and no. Hysteria is, in a sense, something that we *do*, but the distinction between what we do and what happens to us – like the distinction between mind and brain – is less hard and fast than we sometimes assume. The lawcourts require a sharp distinction between willed and involuntary actions. But there is evidence that people can find it hard to distinguish actions they cause from movements that are caused. And, of course, as the American philosopher Thomas Nagel has put it, 'To do anything we must first be something' – and that something is not of our choosing. Weir Mitchell surely was right: 'The largest knowledge finds the largest excuses . . . . The elements out of which these disorders arise are deeply human, and exist in all of us, in varying amounts.'

Hysteria guides us out of the orthodox realm of disease, through which we have been travelling in the previous chapters of this book, into what is, for most of us, a much more familiar domain – one brimming over with the 'human concerns' that first attracted Freud to the study of medicine. This is, if you will, the domain of the soul. In the final chapter I will ask how this territory of the spirit connects with the sphere of the brain.

CHAPTER 10

Soul

The Anatomy of the Soul

Soul! The great, most meaningful word! . . . the very being of the self.
How they drag you down by making you the slave of the body!
Johann Heinroth, *Psychologie*

With insertion into the human individual an immortal soul . . .
a trespass is committed. . . . It is an irrational blow at the
solidarity of the individual.
Sherrington, *Man on His Nature*

THE HUMAN MACHINE

Our ancestors . . . were machines
(made of machines made of machines). . .
Daniel Dennett, *Kinds of Minds*

Where have we come from?

Soon after the birth of the universe, fourteen thousand million years ago, there were no atoms and no elements – only energy and elementary particles, streaming away from the unimaginable explosion that began the history of time. As temperatures fell, the subatomic particles that form the simplest atoms, of hydrogen and helium, coalesced. Matter attracting matter, the earliest stars and galaxies were formed. In the burning centres of these stars, lighter atoms fused to create the heavier elements of which you and I substantially consist – like carbon and oxygen, nitrogen and phosphorus, calcium and iron. Oxygen deficiency was, you may recall, the cause of Alison's gradual decline into coma in Chapter 1.

Ten thousand million years later, soon after the formation of the earth, complex molecules, like DNA, capable of reproducing themselves, formed on our young planet. Through a process of mutation and selection, they evolved, giving rise within a thousand million years to organisms roughly akin to our bacteria. These mini-creatures lacked some of the niceties of civilisation, but were already equipped with much of what it takes to make a man – or a mouse: genes, proteins, a host of other 'macromolecules' and the intricately regulated set of chemical reactions required to sustain life. The smallest conceivable abnormality of a single gene in Charley's DNA preordained his compulsive restlessness of mind and body. Pete, and his small group of fellow victims of 'new variant CJD', succumbed to attack from one of our biology's cruel, unexpected tricks, when against all expectation a protein became fatally infectious.

Somewhere around two thousand million years ago, the ancestor of our modern, compartmentalised cells emerged. This came about when two of their more primitive predecessors embarked, as we saw in 'Metamorphoses', on the most intimate of coalitions, one entering the other permanently. About a thousand million years ago, groups of these more sophisticated cells combined, creating organisms with many cells. One of these, our ancestor, the fish, was swimming in Cambrian seas five hundred million years ago. By then, they and other animals were making use of highly specialised cells to coordinate their activities in the light of their past and present circumstances. These cells were neurons, the basic components from which the nervous systems of all animals on earth were, and still are, built. They are, of course, also the building blocks of our brains. Dave's moments of epiphany resulted from the failure of a small contingent of these cells to find their destination in his growing brain. Lucy's compulsive naps, bad nights and scary dreams were caused by the loss of a single neurochemical that normally signals between the neurons that control our sleep and waking.

One promising evolutionary path – though by no means the only successful one available – involved the gradual sophistication of nervous systems, allowing their possessors to make ever finer perceptual distinctions, and engage in ever more complex forms of behaviour. Our mammalian ancestors took this path, and became the intellectuals of creation. In the biggest and best mammalian brains, like those of monkeys, a growing proportion of neurons were buried in the networks that performed the computations on which intelligence depended: relatively few now sensed or acted directly on the world. Over the past five million years, since the evolutionary paths of man and ape diverged, the pace of change quickened dramatically. Brain weight among our hominid ancestors climbed from around 500 to around 1,500 grams, with correspondingly rapid gains in intelligence. Our cortex had to fold up tight to squeeze inside our heads. These rapid evolutionary advances opened up new spaces for disorder, like Shona's protracted *déjà vu*, Jed's transient amnesia, Jan's progressive loss of knowledge of the world – and Jenny's 'romance of morals and manners'.

This is the briefest of sketches of the evolution of the human brain. Its organisation is more complex than any other known system in the universe: yet its basic ingredients are relatively simple – and Hippocrates was absolutely right: 'To know the nature of man we must know the nature of all things.' The previous chapters have shown how atom, gene, protein, organelle, neuron, neurotransmitter, neural network, lobe and hemisphere must all be invoked, in their place, to understand the functions and dysfunctions of the brain.

But the brain is more than the sum of these parts. It is also – surely – the organ of mind, the seat of consciousness, the wellspring of the soul. The time has come to focus on these higher functions of the brain, to ask how events in our brains give rise to the life in our minds. This closing chapter explores some tentative answers to this ancient question. I suspect that, as we normally pose it, the question is simply unanswerable, that somehow or other we are asking the wrong kind of question – but finding our way to the right one is hard.

THE HISTORY OF IDEAS

This book is an introduction to the science of the brain. This science is exciting partly because it connects – to a greater extent than, say, the science of the heart or the kidney – with so much of the rest of life: 'neurology is the study of mankind itself'. But nevertheless the science of the brain remains just that, a science. Like every science it works with concepts that have been carefully defined in terms of observations that anyone, with the right expertise and equipment, can make – concepts like neuron and synapse.

The words that we use when we think about the mind are quite different. 'Mind', 'soul', 'consciousness' are not scientific terms, and lack technical definition. Our understanding of them is powerfully influenced by religious and philosophical traditions. So there is an inevitable risk of a disconnection here between scientific enquiry and everyday thought.

These three ideas are linked, but have subtly different connotations. The soul – the supposed immaterial, invisible, immortal element that animates our being – clearly has religious overtones. The mind is its secular counterpart: we use the word to refer to the set of abilities that underlies our thoughts and emotions, personality and behaviour. 'Consciousness' can refer to a state of arousal – wakefulness – but also to whatever may be passing through our mind at a given time, our experience or awareness.

I hold out least hope for a scientific explanation of the soul, for reasons I explain in the next section. The sciences of mind and consciousness are thriving, but many thoughtful commentators remain doubtful that science can provide a complete understanding of either. The rest of this chapter will explore the reasons why – and ask whether they are right.

IMAGINING THE SOUL

What sets us apart is a life in the mind, the capacity to imagine.
Robin Dunbar, *The Human Story*

We are imaginative apes: inveterate daydreamers, night-dreamers, tellers and hearers of tales; we imagine events from the past, the faces of friends, our favourite places, a myriad futures; we are captivated by the thought that ours is but one among countless possible worlds, delighting in such moments as when – in C.S. Lewis's famous children's tale – a small girl steps through a musty cupboard into the icy crispness of Narnia's winter. This capacity to detach ourselves from the here and now, to journey mentally into the past and future, into fictional lives and previously unimagined lands, into subatomic spaces and intergalactic space, is surely, as the primatologist Robin Dunbar suggests, the defining mark of man.

What we imagine sometimes comes to pass. We imagined that we might land men on the moon, and achieved this outrageous ambition. I imagine that I may finish writing this book soon. But much that we imagine stays that way: I never plucked up courage to introduce myself to the tall, pale redhead who haunted my thoughts for a week or two twenty years ago – though I vividly imagined doing so. And we sometimes imagine pretty much *impossible* things: there are, I'm afraid, no kittens fluent in Welsh, no statues carved from stone that breathe, no carpets that transport their owners through the air even when asked most politely to do so.

Some kinds of imagining arise more naturally than others. Discovering that the earth was once populated by the dinosaurs, or the intricacies of DNA, required huge efforts of imagination, but relatively few of us are temperamentally inclined to delve into these obscure corners of the natural world. Almost all of us, on the other hand, devote plenty of time to imagining each other – and each other's minds.

There is a powerful human tendency, apparent across cultures and historical time, to imagine the following, quite remarkable, things about the mind: that it can be prised apart from the body; that it can survive the body's death; that it can – on occasion – foresee the future, act directly on physical objects and communicate directly with other minds; that it is bound, by a moral contract, to honour the laws established by its creator. These ideas may seem entirely natural to you, or quite outlandish, but they have undoubtedly been accepted far more often than they have been rejected by human societies. Imagining the mind this way is to imagine it as a 'soul' – the ghostly invisible, immaterial, imperishable twin of our visible, material, self, which is, as we know all too well, fated to return in time to the dust from which it sprang.

Why this undoubted tendency should exist is a fascinating question. The belief in the immortality of an immaterial soul might be explained along these lines. We are used to the occasional absence of those close to us, sometimes in places we have not visited ourselves. It is natural, perhaps, that when someone we love dies, we should imagine them to be absent in some more radically inaccessible place. But we know their body lies in ashes, or beneath the ground. It must therefore be their essential being, their spirit or their soul, that has made the journey to that other place. The beliefs in the wider moral purpose of natural events, and a moral contract that involves us in that purpose, may be explained by our almost insuperable tendency to look for human meanings in the world, to anthropomorphise our environment, a tendency that can in turn be traced to the supreme importance of human relationships in our intensely social lives. The psychologist Jesse Bering has argued that this family of beliefs – in personal immortality, the higher purpose of the self and the moral significance of natural events – may have been shaped specifically by selective pressures, as it promoted behaviour calculated to preserve 'reputation', with benefits to the individual and the group. It is certain that these beliefs are extremely tenacious. As David Lodge writes: 'the idea of an immaterial self is so deeply ingrained in our language and our habits of thought, whether or not we are religious believers that it seems to me doubtful that it will ever be expunged'.

I, like many of us, am excited by these supernatural ideas linked to the core idea of the soul – of journeys between the worlds, foreknowledge of the future, telekinesis and the rest. They give me satisfying shivers down the spine when I encounter them in literature,

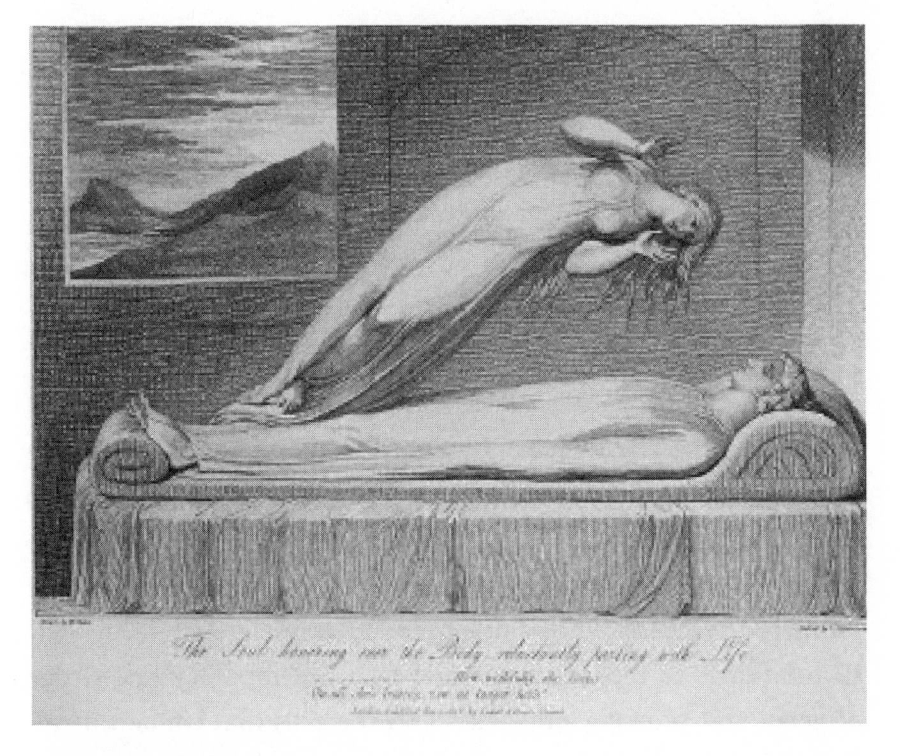

10. The Soul Departing
Blake's figure shows 'The soul hovering over the body reluctantly parting with life'. Engraving by Louis Schiavonetti (1765–1810) after Blake.

where they are so often explored: Charles Dickens, C.S. Lewis, J.K. Rowling, Philip Pullman, for example, engage repeatedly with these themes. But to which class of 'imaginings' do these beliefs belong? The imaginings that we know have come to be, like moon flight; those that could be or might have been true, like my date with the redhead; or the ones we are firmly convinced are untrue, like the tale of the Welsh-speaking kitten? I owe it to you to confess: I am firmly convinced that the comforting and tenacious idea that we possess an invisible, immaterial, imperishable soul is no more than a wonderful fiction.

If we reject this comforting idea as a mythology, an appealing fiction in which we can no longer seriously believe, what becomes of our view of the mind? Is all that remains a loveless prospect of 'atoms and the void'? Perhaps the fate of another great mystery, the mystery of life itself, has something to teach us about the mind.

MATTER, LIFE, MIND

In the nineteenth century, the existence of life was regarded by many scientists, including Pasteur, the great French microbiologist, as an unfathomable mystery, the expression of *élan vital*, the irreducible essence of life. The discoveries we touched on in the first few chapters of this book make this hypothesis unnecessary. Life simply *is* the set of processes that allows organisms to utilise energy from their surroundings to reproduce themselves. Contemporary knowledge of genes and proteins, organelles and the cells that contain them, shows that these goals can be achieved by natural, physicochemical processes, with no need to invoke any further, mysterious, force. Life is not one unfathomable thing: it is the coherent operation of a great many intricate but fathomable things.

Can we follow this line of thought through to the mind? Just as life arises, by natural processes, from the workings of organised matter, might not mind emerge, by natural processes, from the workings of organised life? The idea is surely appealing. Consider two of the hallmarks of mind – knowledge and intelligent behaviour. It is easy to see that gaining knowledge of the world, and using it to guide behaviour, could be advantageous for an organism, enhancing its ability to adapt itself to its environment. In due course, it might become useful for it to gain some knowledge about itself, and about its own knowledge-gaining system, so that it comes to model its own mind. We know now that the system we actually use for all these knowledge-gaining and knowledge-using purposes is the brain. It is immensely tempting to suggest that, roughly speaking, mind stands in the same relation to the workings of the brain, and the activities of the body it controls, as life stands to processes of reproduction and metabolism.

We have encountered some of these knowledge-wielding structures in previous chapters – the network of networks in the limbic system that allows us to form 'episodic' memories of particular events, the regions of the left temporal lobe that store our accumulated 'semantic' understanding of the world. I could have given similar brain-based accounts of each of the other components of 'cognition', the collective term for the aspects of mind involved in gathering, storing and using knowledge: attention, perception, mental control, language, praxis. The exploration of the brain is beginning to make scientific sense of the mind.

Can the same be said of consciousness – our awareness, our experience? Here, too, there has been plenty of progress. Much of what happens in the brain happens unconsciously – the regulation of hormones by the hypothalamus, the control of breathing by the brain stem, even the control of visually guided movement by parts of the cerebral cortex, can all work just fine in the absence of conscious awareness. So what is it that distinguishes brain activity that is associated with awareness?

There are several candidates, with some support for each. Quantity seems to matter: small amounts of activity are less likely to enter consciousness than large ones. Quality counts too: coherent activity, synchronised across collaborating neurons, stands a better chance than less organised activity. Location may be important: in general, activity in the cortex is regarded as being more likely to enter awareness than activity elsewhere. But recent work has particularly emphasised the importance of a fourth characteristic, the 'reach' or connectivity of the activity. The underlying idea is that consciousness depends on communication between several, usually widely separated, regions. According to the 'global workspace theory', proposed by Bernard Baars and Stan Dehaene, the crucial distinguishing feature of the information in consciousness, shaping our awareness, is that it is broadcast widely through the brain, gaining access, as it does so, to the neural processes that control our actions and allow us to report on our experience.

Despite our new understanding of biology, life remains creative, unpredictable, wonderful – but, given this understanding, it no longer strikes us as an unfathomable mystery. As we learn more and more about the processes in the brain that underpin it, will we similarly lose our sense of mystery about the mind?

WINE AND WATER

To man's understanding the world remains obstinately double.
Sir Charles Sherrington, *Man on His Nature*

Some authors, like the trenchantly materialistic philosopher Daniel Dennett, believe so strongly. But others distrust the analogy I have drawn between life and mind. Though he is not the most recent of these doubters, the misgivings singled out by Sir Charles Sherrington, the English physiologist, in *Man on his Nature*, written in 1937, late in his long life, carry particular weight. Sherrington was a brilliant physiologist and desperately eager to make biological sense of the mind. He felt that the failure to do so was 'embarrassing for biology'. The trouble was that he just couldn't see a way to do it.

Like contemporary doubters, Sherrington accepted that *life* could be understood in physicochemical terms, but the nature of mind remained for him an agonising problem.

His key misgiving about the possibility of a scientific understanding of the mind was that the data in question, the contents of awareness, were, he believed, simply invisible to the scientist: 'Our mental experience is not open to observation through any sense organ. . . . The perceptible [is] rooted in sense. Our mental experience has no such channel of entrance to the mind. It is already of the mind.' Given that the scientist could never hope to find what he was looking for – experience – in the brain, but only neurons, in their cosmic numbers, interlinked and ceaselessly communicating, Sherrington came to the conclusion that 'To man's understanding the world remains obstinately double' – a permanent double act of matter-life and mind. Science, he reluctantly concluded, could describe the workings of the brain, and the behaviour that they drive but experience, the essence of the conscious mind, involved a set of 'further facts', fated to lie for ever beyond the reach of scientific study.

Several eloquent recent thinkers have broadly agreed with Sir Charles. In his now famous essay, 'What is it like to be a bat?' the American philosopher Thomas Nagel argues that no amount of scientific exploration of a sense we lack – like echolocation – will ever give us a feel for 'what it's like' to have and use that sense. A certain kind of subjective, first-person knowledge seems to lie beyond the reach of science. The philosopher Colin McGinn summarised the problem pithily: 'The water of the physical brain is turned into the wine of consciousness, but we draw a total blank on the nature of this conversion . . . . There is something terminal about our perplexity.' David Chalmers, a highly imaginative current thinker in the field, introduced a related distinction that has been taken up widely since. The 'easy questions' of consciousness concern its mechanisms, the kinds of questions I was beginning to answer in the last section, about what sorts of process in which parts of the brain are required for awareness. The 'hard' question is why these processes should give rise to consciousnesss *at all.* To many the hard question looks unanswerable.

The gauntlet flung down by Sherrington and his philosophical followers offers a serious challenge to the science of the mind. But it may be that the challenge – inescapable as it seems – is somehow the wrong one, and that we should leave the gauntlet where it lies. How could this be?

Sometimes, when we're perplexed, in a dilemma, we have to trust to the stronger of two conflicting intuitions, particularly if we have a hunch that one of the two is flawed. For me, though clearly not for everyone, the intuition that mind is continuous with life, just as life is continuous with matter, is very strong. I suspect that the opposing, 'dualistic', intuition, that mind-stuff or mind-properties are of a different order from material ones, originates at least partly in the tenacious supernatural conception of the soul. Even people who long ago abandoned any formal faith continue to be influenced by an intellectual history in which religion has been immensely formative. Abandoning the soul will not come easily.

Its abandonment may be eased just a little by a second reflection. The problem we are travelling around, of the relation of mind to matter, tends to presuppose a polar distinction between the two, something like the radical contrast Descartes drew between 'res extensa' and 'res cogitans', extended things – objects – and thinking things – minds. If we give up Descartes' strange conception of mind cut free from matter, relinquishing the concept of the soul, we need not feel committed either to Descartes' conception of matter free of mind. The classical problem of mind and matter begins here – with the opposition of mindless matter to matterless mind. But arguably this opposition is false. The relationship of mind and matter is not oppositional but intimate and circular: mind emerges from matter, and matter is conceived by mind. The polar opposition of subject and object is also misleading, for the same reason. We can never attain complete objectivity: knowledge is always shaped by the knower. But nor will we ever enjoy pure subjectivity: knowledge always arises from and contains something of the world.

These thoughts suggest an answer to the question I posed earlier: if the soul is a fiction, are we left with no alternative in our understanding of ourselves but a bleak materialism? I think not. There is no need to 'find a place for mind' in a physical world: the mind is in the picture from the start. *We* are in and of the world, from our beginning.

And *we*, of course, are more than brains. Perhaps another reason why we find ourselves asking an impossible question about the relationship of mind to matter is that in our exclusive focus on the brain we have shut out the materials from which we might have built a serviceable answer.

FROM THE BRAIN ONLY?

From the brain only arise our pleasures, joys, laughter and jests,
as well as our sorrows, pains, griefs and tears. . . .
These things that we suffer all come from the brain.
Hippocrates, *On the Sacred Disease*

We tend to regard the brain as the big boss, the chief executive, the guy in charge. It is, appropriately, installed in the human penthouse, and has the best view that our fragile frame can provide. Like its counterparts in business, it is exceptionally well paid, consuming much more energy for its weight, when we're at rest, that any other organ in the body. We think of it as a symbol of the soul – or, at least, as its material embodiment. I spend most of my waking and working hours puzzling over the ingenious things it does, and I am loath to belittle its importance. But we may be at risk of a misleadingly 'neurocentric' view of things. If we hope to discover all the richness of experience in the workings of the brain alone, we risk serious disappointment.

Imagine you are standing on a headland, looking out to sea. A hundred feet below you, countless waves are pounding in, crested by white foam. Your eyes travel over league after league of boisterous water, to find the horizon where a turquoise sea encounters a pale blue sky. What could be simpler than opening your eyes to see this view? Well, consider the prerequisites for this elemental experience.

You must, first, most fundamentally, be alive – but not simply alive: you need to be the kind of living thing that moves (and fidgets), the kind that needs to find its way around, for otherwise, why go to all the trouble it takes to see in the first place? You have to be, in other words, an animal, one of those 'fires that burns without light', a needy, breathing, palpitating, reproducing creature. Such creatures, over several hundred million years, evolved a delicate instrument with which to detect light, and distinguish its wavelengths, form an image of the world and transmit the news of their arrival. Darwin came close to admitting defeat when confronted by the exquisite mechanism of the eye, writing that 'to this day it gives me a cold shudder'. But the eye, of course, did not evolve alone. Evolution operates by way of reproductive outcomes that eyes can't achieve in isolation – eyes are useful only when they have bodies to guide. Vision was embodied from the start, and every part of the visual system, from eye through the brain to the hand that it guides, has been shaped by natural selection. The visual system already contains a huge amount of information about the world it is designed to see before we ever open our eyes.

Coming by an eye – and even a visual system – is just a beginning. Your view of those breaking waves is also the outcome of a long-forgotten process of development. Over the first few years of infancy and childhood, during the period of maximum plasticity in the nervous system, the cells in our visual system are gradually tuned to features in the world: our brains are literally shaped by our environments. At the same time, as we learn how to find our way about the world using our eyes, we build up a huge stock of expectations about how the appearances of things in the world will change as we move. If I look at a vertical line and move my eyes a little to the side, it blurs. Knowing this is part of what it is to see the line. People blind from their early years who have sight restored in adulthood complain that they 'can't see', despite the undoubted flood of information streaming in from their eyes. They lack the stock of expectations that usually enables us to negotiate the visual world without apparent effort. As the astronomer Sir William Herschel wrote: 'seeing is an art that must be learned: we cannot see at sight'. Such 'sensorimotor' knowledge grounds all our later cognition.

All the while, as human infants gaining perceptual knowledge, we are exposed to language. We map it gradually on to our growing practical understanding of the world. Language provides a rapid entrée into the cultural realm that soon begins to pervade and shape experience: the sea we look out on becomes the taking-off point for a thousand voyages – real and imagined, of adventure, endurance, exile – that hover around the headland view.

There is more, yet, to our seeing. As we look out over the breaking waves, our eyes are in motion. They interrogate the view. Experience is a process of skilful intelligence-gathering rather than passive reception. To perceive is always to act.

However simple an experience seems at first glance, it is always 'embodied, embedded, and extended' – the activity of an embodied creature, embedded in a culture, engaging in an interaction with surroundings and other people which extends over space and time. The brain is a vital link in the chain of experience, but not its sole source, its only begetter. We tend to think of it as a kind of magic lantern – once rubbed the right way it gives rise to a stream of invisible, immaterial 'mind events' to parallel the visible, material stream of brain events. Perhaps we should think of it instead as an enabler – an immensely subtle instrument that brings us into contact with the world, enabling us to apprehend its extraordinary richness.

But I may have left you feeling cheated. For if this is all that science can offer, a description of events occurring in a highly evolved and educated brain hooked up to a body moving through space and time, has it not failed to capture the essence of experience, the feeling of what it's like to be alive?

THE KEY OF THE KINGDOM

In that kingdom is a city,
In that city is a town,
In that town there is a street,
In that street there winds a lane,
In that lane there is a yard,
In that yard there is a house,
In that house there waits a room,

In that room there is a bed,
On that bed there is a basket,
A basket of flowers.

Flowers in the basket,
Basket on the bed,
Bed in the chamber,
Chamber in the house,
House in the weedy yard,
Yard in the winding lane,
Lane in the broad street,
Street in the broad town,
Town in the city,
City in the kingdom:
This is the key of the kingdom.

Traditional

This nursery rhyme echoes a game that some of us played as children, zooming in and zooming out of our location. As it swoops from the grand to the domestic, from the scale of nation to the scale of home and heart, and then reverses the procedure, it seems to imply that we need to take into account *all* the resulting perspectives to make proper sense of things – to unlock the kingdom. It offers an analogy to the journey travelled in this book, from the minute atoms of our being to the visible expanses of the brain. But it can also be read as a metaphor for the contrast between two ways of thinking about experience – through science and through art.

By its nature our experience, our consciousness, is unique – the preserve of a particular subject, the individual product of our own, highly personal, history. It represents a massive synthesis of a vast amount of information, the outcome of an unparalleled process of integration. We – and our brains – are the synthesisers, the integrators that make such a thing possible. Although this achievement is complicated, it is not intrinsically mysterious: but it is a vital fact about our nature. Capturing and evoking the 'dense specificity of personal experience', in the words of David Lodge, is one of the primary aims of art – *the* primary aim, perhaps. Lodge quotes Joseph Conrad, from the preface to one of his tales: 'My task which I am trying to achieve is by the power of the written word to make you hear, to make you feel – it is before all to make you *see*. That – and no more, and it is everything.' Art aims to conjure up the living presence of experience: as David Lodge has argued, artists are, in this sense, the true experts on consciousness.

Science has a different remit. Even when it is trying to explain experience itself, it looks beyond it, to the principles and regularities that govern its behaviour. It does not aim to conjure up its living 'presence': instead, it provides a 'likeness' to explain – but not evoke – what we experience. Instead of synthesis and integration, it works through analysis and differentiation. Where art is particular, personal, specific – even if of universal relevance – science strives to be general, impersonal, abstract. Each has much to say about experience: the approaches are complementary. Both offer keys to the kingdom, helping us to understand our minds and lives.

EPILOGUE

O Magnum Mysterium

O magnum mysterium, et admirabile sacramentum . . .
(O great mystery, much to be marvelled at . . .)

Some pieces of music are best listened to alone – so intoxicatingly beautiful that the pleasure calls for absolute absorption. I first heard Morten Lauridsen's setting of 'O Magnum Mysterium' on the radio, driving to work, and almost crashed. How could any group of human voices create a music so lovely – or so sad?

It celebrates a mystery – the mystery of a fateful birth, in the midst of animals, cast out in a manger. I am not a religious person, but this music summons up a tenderness and sense of awe, a tearful mix of wonder, happiness and grief. I seek it out from time to time. I mention it now because the power of music to command our feelings is one of the best illustrations I know of the principal theme of this book: the deep continuity between matter, life and mind.

How does this strangely abstract art cause us to catch our breath and weep? The clue lies, surely, in the word we so often use of music – that it *moves* us. No other form of art speaks so directly to our feelings, but music speaks just as directly to our tapping feet and dancing bodies. It keys readily and powerfully into the brain's activity. In a study of the brain's response to the particular, personally chosen, music that reliably evoked shivers down the spine in a group of musicians, Anne Blood and Robert Zatorre found that the shivers were accompanied by a reliable rise in heart rate, depth of breathing, muscle tone – and more: a pattern of activity in the brain, in regions linked to arousal, reward and emotion, that mirrors the changes occurring in response to natural stimuli, like food and sex, that can induce euphoria. In a complementary case report entitled 'When the feeling's gone', the neurologist Tim Griffiths described a patient whose intense emotional response to the Rachmaninov preludes – but not his capacity to judge rhythm or melody – was abolished by a stroke affecting some of the areas identified by Blood and Zatorre.

Music can capture the very form of feeling, a form that is at once mathematical and mindful, echoing the motions of particle, and planet, and the dance within our brains. Music is so enrapturing because, in a deep sense, we, ourselves, *are* music – not least in the straightforward sense that our brains are governed by a constant interplay of rhythms. To borrow Oliver Sacks's memorable phrase, music gives us a chance to glimpse 'the quantitative and the qualitative fuse' – as the mathematical pattern of notes on the page is translated into a pattern of movement in the musician's arm or singer's throat, that sets up a pattern of movement in the air, that gives rise to a pattern of signalling in a listener's brain, that becomes a pattern of movement tapped by his feet, felt in his mind.

We are only beginning to understand how such things can happen – how the lifeless gives rise to the living, how the human 'quintessence of dust' can be 'in action . . . like an angel! In apprehension . . . like a god!' The answer outlined in the course of this book is a tentative sketch. We are intensely colonial creatures, colonial at multiple levels. We are made of atoms: whatever laws govern the atom, govern our lives. Our atoms combine in giant molecules, like proteins and genes, exquisitely fitted for their purpose by three thousand million years of terrestrial evolution. These molecules themselves join forces to create the organelles, like mitochondria, that conduct the biochemical business of life in our cells, deriving energy from the environment to build and maintain themselves. Cells are organised into tissues: the tissues of the brain are an intricate tapestry of networks of neurons, and networks of networks of neurons, designed for command and control. We really are, as Daniel Dennett puts it, made of 'machines made of machines made of machines': machines designed to replicate themselves. Yet we are something other: we have minds. Mind is not a different kind of machine – invisible, imperishable, immaterial. It is one of the activities of life, a crucial one, not least because it allows us to come to know matter. Mind is at home with matter. It arises from matter and it conceives it: it is both its child and its parent.

In the nightmarish tale of Dr Jekyll and Mr Hyde, Robert Louis Stevenson darkly predicted that 'man will ultimately be known for a mere polity of multifarious, incongruous and independent denizens'. This book has introduced you to some of the distinctive denizens of man, and woman, the inner parts and humming networks of the neurons of the brain. You may feel anxious for your self, your soul, threatened by disintegration into an ever-lengthening series of alien components, 'millions of local principles'. I sympathise. But, somehow, as a rule, the parts add up. Like it or not, they are the fabric of our being. We are learning to make sense of ourselves without invoking supernatural powers. 'There is grandeur in this view of life,' as Darwin wrote. There is also mystery aplenty.

Sources and Suggestions for Further Reading

BACKGROUND

Several collections of neurological tales ('a sort of *Arabian Nights* entertainment', in Sir William Osler's words) will appeal to readers who enjoy this book. These include Oliver Sacks's wonderful *The Man Who Mistook his Wife for a Hat* (Duckworth, 1985) and his subsequent works, including *An Anthropologist on Mars* (Picador, 1995). Paul Broks's *Into the Silent Land: travels in neuropsychology* (Atlantic Books, 2004), does for neuropsychology what Sacks did for neurology, and heads off into wilder imaginative realms. Jonathan Cole's *Pride and a Daily Marathon* (MIT Press, 1995) is a detailed but fascinating study of a single neurological case, a man who lost all sensation from his limbs. Harold Klawans's *Newton's Madness* (Headline, 1990) is a quick-fire series of neurological detective stories. Although most of the tales in these books are drawn from life, they have many of the qualities of fiction. Sebastian Faulks's *Human Traces* (Hutchinson, 2005) is an absorbing recent work of fiction, on a grand scale, containing a wealth of information about the history of neuropsychiatry, with concerns that overlap those of all the authors listed here. Stanley Finger's *Minds behind the Brain* –

a history of the pioneers and their discoveries (Oxford, 2000) is an appealing introduction to the wider history of neuroscience.

Those in search of more systematic scientific background might want to delve into one of the current textbooks in the area. Alwyn Lishman's *Organic Psychiatry (*Blackwell, 3rd edn, 1997); Eric Kandel et al.'s *Principles of Neural Science* (Prentice-Hall International, 3rd edn, 1991); Michael Gazzaniga's *Cognitive Neuroscience* (W.W. Norton, 2nd edn, 2002) are my personal favourites.

CHAPTER 1: I AM TIRED

For the history of chemistry, Paul Strathern's *Mendeleyev's Dream: the quest for the elements* (Hamish Hamilton, 2000) and Oliver Sacks's *Uncle Tungsten* (Picador, 2001) are both highly readable and informative. I also consulted Roy Porter's *The Greatest Benefit to Mankind* (HarperCollins, 1997) and W.C. Dampier's *A History of Science* (Cambridge University Press, 1966). Carl Zimmer's entertaining study of Thomas Willis and his contemporaries, *Soul Made Flesh* (Free Press, 2004), and, in more academic colours, Robert Martensen's *The Brain Takes Shape* (Oxford University Press, 2004) touch on aspects of the history summarised in this chapter, especially the scientific revolution of the seventeenth century. Stephen Rose's *Chemistry of Life* (Penguin, 3rd edn, 1991) is an accessible basic introduction to biochemistry (there are many colourful undergraduate textbooks on offer in the area too). The historical background and current state of thought about chronic fatigue syndrome, including a discussion of the many medical disorders that can be mistaken for it, are detailed in S. Wessely, et al.'s scholarly *Chronic Fatigue* (Oxford University Press, 1998). For a brief survey of the same ground see *Essentials of Psychosomatic Medicine*, ed. J.L. Levenson, *Chapter 9*: 'Chronic fatigue and fibromyalgia syndromes', by M. Sharpe and P. O'Malley (American Psychiatric Publishing, 2007). I learned about the dangers of copper deficiency from Stephen Jaiser et al., 'Copper deficiency masquerading as subacute combined degeneration of the cord and myelodysplastic syndrome', *Advances in Clinical Neuroscience and Rehabilitation* (2007): 20–1. Alison's case is described in technical terms in A.S.J. Zeman et al., 'Multicore myopathy presenting in adulthood with respiratory failure', *Muscle & Nerve*, 20 (1997): 367–9.

CHAPTER 2: DON'T FIDGET

Matt Ridley's *Genome: the autobiography of a species in 23 chapters* (Fourth Estate, 1999) is a much-praised popular introduction to genetics. I learned a lot from D.J. Weatherall, *The New Genetics and Clinical Practice*, now in its third edition (Oxford, 1991). Stephen Rose's *Chemistry of Life* again provides some background help. Genetic relationships between the genes of a wide variety of species are discussed in 'Analysis of the genome sequence of the flowering plant *Arabidopsis thaliana*. The Arabidopsis Genome Initiative', *Nature*, 408 (2000): 796–815. Research on the FOXP2 'language gene' is summarised in 'FOXP2 and the neuroanatomy of speech and language', by Faranah Vargha-Khadem et al., *Nature Reviews Neuroscience*, 6 (2005): 131–8. Work relating to the basal ganglia, mood and behaviour includes 'Transient acute depression induced by high-frequency deep-brain stimulation', by B.P. Bejjani et al., *New England Journal of Medicine*, 340 (1999):1476–80, and L. Mallet et al., 'Compulsions, Parkinson's disease, and stimulation', *Lancet*, 360 (2002): 1302–4. Charley's case is reported in A. Zeman et al.,

'McLeod Syndrome: life-long neuropsychiatric disorder due to a novel mutation of the XK gene', *Psychiatric Genetics*, 15 (2005): 291–3.

CHAPTER 3: THE LIGHT OF DAWN

I was first introduced to the extraordinary history of these disorders by the title essay in a collection of riveting stories of medical discovery, *The Ghost Disease*, by Michael Howell and Peter Ford (Penguin, 1986). Much has, of course, been written since. Stanley Prusiner's 'Prion disease of humans and animals', published in a supplement to the *Journal of the Royal College of Physicians of London*, 28 (1994): 1–30, and his Shattuck Lecture – 'Neurodegenerative diseases and prions', *New England Journal of Medicine*, 344 (2001): 1516–26 are both highly informative. The paper describing the discovery of 'A new variant of Creutzfeldt-Jakob disease in the UK', by R.G. Will et al., was published in the *Lancet*, 347 (1996): 921–5. A Ph.D. thesis, Ki-Heung Kim, 'From scrapie to prion disease', University of Edinburgh, 2003, sketches the historical background as does P. Brown and R. Bradley, '1755 and All That: a historical primer of transmissible spongiform encepholopathy', *British Medical Journal*, 1998, 317: 1688–92.

CHAPTER 4: METAMORPHOSES

Metamorphoses are everywhere. Ovid's *Metamorphoses* has recently been translated and transposed in *After Ovid: new metamorphoses*, ed. Michael Hofmann and James Lasdun (Faber, 1994), the source of this chapter's opening quotation. I was introduced by my daughter, who had been wowed by it, to Richard Strauss's *Metamorphosen*, music with a natural gravitas rendered utterly poignant by the knowledge that it was composed after the bombing of the theatres where Strauss had worked and his music had first been performed, in Munich, Vienna and Dresden. The first popular book devoted to mitochondria appeared recently: Nick Lane's *Power, Sex, Suicide: mitochondria and the meaning of life* (Oxford University Press, 2006). Lynne Margulis's *Acquiring Genomes* (Basic Books, 2002) introduces the idea that symbiosis has been a crucial force in evolution. There is a huge medical and scientific literature on mitochondria. Among other sources, I have consulted M.W. Gray et al., 'The origin and early evolution of mitochondria', *Genome Biology*, 2 (2001): 1018.1–1018.5, J. Schmiedel et al., 'Mitochondrial Cytopathies', *Journal of Neurology*, 250 (2003): 267–77, and Patrick F. Chinnery, 'Could it be mitochondrial? When and how to investigate', *Practical Neurology*, 6 (2006): 90–101. Recent work on the role of mitochondria in synaptic growth is summarised by Jane Qiu in 'Fuel for plasticity', *Nature Reviews Neuroscience*, 6 (2005): 93. The MELAS syndrome was described by S. Pavlakis et al., in 'Mitochondrial myopathy, encephalopathy, lactic acidosis and strokelike episodes: a distinctive clinical syndrome', *Annals of Neurology*, 16 (1984): 481–8.

CHAPTER 5: LOST IN TRANSLATION

This chapter draws heavily on Gordon Shepherd's fascinating study of the nineteenth-century exploration of the microscopic anatomy of the nervous system, *Foundations of the Neuron Doctrine* (Oxford University Press, 1991). Santiago Ramón y Cajal's memoir, *Recollections of my Life*, is available in translation by E. Horne Crainie and Juan Cano (MIT Press, 1996). The memoir divides equally into an entertaining description of his rebellious childhood in mid-nineteenth-century Spain and an account of his – in some ways equally rebellious – working life and prolific discoveries. Ecstatic seizures are described in 'Partial epilepsy with ecstatic seizures', by B.A. Hansen and E. Brodtkorb, *Epilepsy and Behaviour*, 4 (2003): 667–73. Dostoevsky's epilepsy is discussed in C.R. Baumann, 'Did Fyodor Mikhailovich Dostoevsky suffer from mesial temporal epilepsy?' *Seizure*, 14 (2005): 324–30. Michael Trimble's recent book, *The Soul in the Brain: the cerebral basis of language, art and belief* (John Hopkins University Press, 2007), includes interesting detailed discussion of the relationship between epilepsy and religious experience. Quotations from Sherrington, here and elsewhere, are from *Man on His Nature* (Cambridge University Press, 1953). Hippocrates' *On the Sacred Disease* (which of course argues *against* the idea that epilepsy has supernatural causes) is available in the Loeb Classical Library series, translated by W.H.S. Jones (Harvard University Press, 1998). For a detailed up to date account of the current state of the subject launched by Golgi and Cajal, see *The Synaptic Organisation of the Brain*, ed. G. Shepherd (5th edn, Oxford University Press, 2004).

CHAPTER 6: DR GELINEAU'S DREAM

I have elaborated my fanciful account of Dr Gelineau from an 'Historical note' by Pierre Passouant in the journal *Sleep*, 'Doctor Gelineau (1828–1906): narcolepsy centennial', *Sleep*, 3 (1981): 241–6. Gelineau's original descriptions 'De la Narcolepsie' were published in the *Gazette des Hôpitaux*, 1880: 626–8 and 635–7. The current state of knowledge of narcolepsy, including the discovery of hypocretin, the neurotransmitter that is lacking in narcolepsy, has been reviewed recently in Y. Dauvilliers et al., 'Narcolepsy with cataplexy', *Lancet*, 369 (207): 499–511.

CHAPTER 7: THE SENSE OF PRE-EXISTENCE

There are several reviews of *déjà vu*, including – briefly – C. Warren-Gash and A. Zeman, '*Déjà vu*', *Practical Neurology*, 3 (2003): 106–9, or more lengthily, A.S. Brown, 'A review of the *déjà vu* experience', *Psychological Bulletin*, 129 (2003): 394–413 and V.M. Neppe's *The Psychology of Déjà vu: have I been here before?* (Witwatersrand University Press, Johannesburg 1983). Literary treatments are discussed in Herman Sno et al., 'Art imitates life: *déjà vu* experiences in prose and poetry', *British Journal of Psychiatry*, 160 (1992): 511–18. The first case of prolonged *déjà vu* mentioned in the chapter was described by Helen Wright et al., in 'Prosopagnosia following non-convulsive status epilepticus associated with a left fusiform gyrus malformation', *Epilepsy and Behaviour*, 9 (2006): 197–203. The concept of *déjà vecu* is introduced in C.J. Moulin, M.A. Conway et al., 'Disordered memory awareness: recollective confabulation in two cases of persistent *déjà vecu*', *Neuropsychologia*, 43(9) (2005): 1362–78. Daniel Schacter's account of memory's failings is *The Seven Sins of Memory: how the brain forgets and remembers* (Houghton Mifflin, 2002). John Hodges discusses the causes of transient amnesia in an excellent short book, *Transient Amnesia* (W.B. Saunders, 1991). Recent work is summarised in Chris Butler and Adam Zeman, 'Syndromes of transient amnesia', *Advances in Clinical Neuroscience and Rehabilitation*, 6 (2006): 13–14. Jed's specific form of transient amnesia is explored in detail in C. Butler et al., 'The syndrome of transient epileptic amnesia', *Annals of*

Neurology, 61 (2007): 587–98. The 'sleeping and forgetting' section title is borrowed from B.J. Corridan et al., 'A case of sleeping and forgetting', *Lancet*, 357 (2001): 524. The work of Faraneh Vargha-Khadem on the effect on memory of early hippocampal damage is described in 'Differential effects of early hippocampal pathology on episodic and semantic memory', *Science*, 277 (1997): 376–80 and in several more recent publications. For facts about synaptic numbers in complex nervous systems, see Peter R. Huttenlocher's *Neural Plasticity* (Harvard University Press, 2002). For a detailed account of the organisation of the brain's major neuronal systems – the culmination of the work begun by Santiago Ramón y Cajal – see Gordon Shepherd's book mentioned above, *The Synaptic Organisation of the Brain*. Broca's seminal work on the localisation of speech in the left hemisphere was translated into English in 'Translation of Broca's 1865 report: localization of speech in the third left frontal convolution', by E.A. Berker et al., *Archives of Neurology*, 43 (1986): 1065–72.

CHAPTER 8: THE ART OF LOSING

The classical recent description of 'semantic dementia', Jan's disorder, is by John Hodges et al., 'Semantic dementia: progressive fluent aphasia with temporal lobe atrophy', *Brain*, 115 (1992): 1783–806. Semantic dementia is one of the syndromes of fronto-temporal dementia: *Frontotemporal Dementia Syndromes*, ed. John Hodges (Cambridge University Press, 2007), will give an authoritative description of the wider picture. The release of creativity that occasionally occurs in the course of these disorders is described in B. Miller et al., 'Emergence of artistic talent in frontotemporal dementia', *Neurology*, 51 (1998): 978–82 and 'Functional correlates of musical and visual ability in frontotemporal dementia', *British Journal of Psychiatry*, 176 (2000): 458–63. Narinder Kapur discusses related ideas in 'Paradoxical functional facilitation in brain-behaviour research', *Brain*, 119 (1996): 1775–90. Chris McManus's *Right Hand, Left Hand* (Phoenix, 2002) explores the topic of asymmetry. For the view that schizophrenia is related to a failure of hemispheric specialisation, see T.J. Crow, 'Cerebral asymmetry and the lateralisation of language: core deficits in schizophrenia as pointers to the genetic predisposition', *Current Opinion in Psychiatry*, 17 (2004): 97–106. Anyone keen to know more about Rembrandt will have a wonderful time with *Rembrandt's Eyes* by Simon Schama (Penguin, 1999).

CHAPTER 9: BETRAYAL

Ilze Veith's *Hysteria: the history of a disease* (University of Chicago Press, 1965) is a fascinating introduction to thinking about hysteria over the centuries. Recent studies include *Contemporary Approaches to the Study of Hysteria: clinical and theoretical perspectives*, ed. Peter Halligan, Chris Bass and John Marshall (Oxford University Press, 2001). The evidence that around a third of patients attending neurology clinics have medically unexplained symptoms is to be found in A.J. Carson et al., 'Do medically unexplained symptoms matter? A prospective cohort study of 300 new referrals to neurology outpatient clinics', *Journal of Neurology, Neurosurgery and Psychiatry*, 68 (2000): 207–10. Jon Stone has written widely in the area, though many of his key findings, from a study of over 100 patients with symptoms similar to Jenny's ('Functional weakness', Ph.D. thesis, University of Edinburgh, 2006), have yet to be published. His experience, widely shared, is that reports of the death of hysteria have been greatly exaggerated.

CHAPTER 10: THE ANATOMY OF THE SOUL

Heinroth's remark about the soul is quoted in *A History of Medical Psychology*, ed. G. Zilboorg (W.W. Norton, 1941). Sherrington's comes from *Man on His Nature*: Sherrington's view of the relationship of mind to brain was complex, but he had no truck with the idea of the soul. Dan Dennett's *Kinds of Mind: towards an understanding of consciousness* (Phoenix, 1996), is an entertaining introduction to his challenging thoughts on the subject. Robin Dunbar's *The Human Story* (Faber, 2004) introduces the facts of human evolution succinctly and elegantly. Jesse Bering's paper, 'The folk psychology of souls', *Behavioural and Brain Sciences* (Cambridge University Press, 2006) discusses possible evolutionary advantages of believing in the existence of soul and afterlife. David Lodge's *Consciousness and the Novel* (Penguin, 2003) takes an oblique, thought-provoking look at the science of consciousness, arguing the case that the real experts on consciousness are, in fact, creative artists. The distinction between 'likeness' and 'presence' is developed along interesting lines in Robert Martensen's *The Brain Takes Shape* (Oxford University Press, 2004). My own *Consciousness: a user's guide* (Yale University Press, 2002) explores this chapter's territory in more detail: its references would guide you to other writings in the area. I have tried to move just a little way beyond the agnostic view I took there in this chapter, pursuing the thought that mind is 'embodied, embedded and extended'. Douglas Hofstadter's *I Am a Strange Loop* (Basic Books, 2007) is a wonderful study of the soul (and self, interiority, consciousness), taking off pretty much where this book stops: I strongly recommend it to anyone who wants to explore ways of making sense of the mind, without invoking magic, on the basis of our current understanding of the brain. My current reading (unfinished) in this eternally interesting and equally perplexing line of country is George Lakoff and Mark Johnson's *Philosophy in the Flesh: the embodied mind and its challenge to Western thought* (Basic Books, 1999) and M.R. Bennett and P.M.S. Hacker's *The Philosophical Foundations of the Neurosciences* (Blackwell, 2006). 'The key of the kingdom' is a nursery rhyme, which can be found in Iona and Peter Opie, *The Oxford Nursery Rhyme Book* (Oxford, 1955).

EPILOGUE: O MAGNUM MYSTERIUM

Anne Blood and Robert Zatorre's study of shivers down the spine is: 'Intensely pleasurable responses to music correlate with activity in brain regions implicated in reward and emotion', *Proceedings of the National Academy of Sciences*, 98 (2001): 11818–23. Tim Griffiths's case report is: 'When the feeling's gone: a selective loss of musical emotion', *Journal of Neurology, Neurosurgery and Psychiatry*, 75 (2004): 344–5.

Glossary

This glossary provides brief reminders about the meanings of the more technical terms used more than once in the course of the book. Words used only once and explained in their context are not always included here. Where a word plays an important part in a single chapter, the chapter in question is mentioned. Words in italics are themselves defined elsewhere in the glossary.

Acanthocyte
A red blood cell with thorny protuberances deforming the *cell*'s normally smooth doughnut-like shape. These cells are the signature of a group of neurological disorders, known collectively as 'neuroacanthocytosis', affecting movement, thought and behaviour. *McLeod's syndrome* (Ch. 2) is a member of this group.

Acetylcholine
A chemical which transmits excitation between nerve and muscle. It is also a major *neurotransmitter* in the brain, with an important role in maintaining arousal and permitting learning. Brain acetylcholine is depleted early in the course of *Alzheimer's disease.*

Action potential
The all-or-nothing electrical signal transmitted by a *neuron* that has been sufficiently excited (Ch. 5).

Alpha rhythm
An electrical rhythm at 8–13 cycles/second originating in the brain which can be recorded from the back of the head in a relaxed subject with his eyes closed.

Alzheimer's disease
The commonest form of *dementia*, usually causing memory loss for recent events, in the first instance, with gradual involvement of other intellectual abilities, and related effects on mood, personality and behaviour (Ch. 8).

Amino acid
Amino acids are the building blocks of *proteins*. Our bodies make use of about twenty. The genetic code operates by specifying the order in which amino acids are strung together to build proteins. Amino acids also serve other biological functions. For example, the amino acid *glycine* doubles up as a *neurotransmitter*, and the amino acid *glutamate*, itself an excitatory neurotransmitter, is the chemical parent of the inhibitory neurotransmitter gamma-aminobutyric acid (*GABA*) (Ch. 3).

Amnesia
Inability to form or recall new *memories*. In human sufferers the defect is usually selective, for example affecting memory for events but sparing the ability to acquire new *motor* skills. The classic amnesic syndrome, causing inability to form new memories with a variable loss of memories for the past, is due to structures linked within the *limbic system*, especially the *hippocampus* and *thalamus*. Amnesia can also occur transiently (Chs 7, 8).

Asymmetry
In *neurology*, this refers to the specialisation of function of the two sides of the brain, the left hemisphere taking the leading role in language processing, the right in perception (Ch. 8).

Atom
Atoms are the building blocks of elements: an atom is the smallest possible quantity of an *element*, such as oxygen, hydrogen, sodium or gold. Atoms of the same or different elements can combine to form *molecules* (a water molecule consists of two hydrogen atoms and one oxygen atom, H_2O – see Ch. 1).

ATP
Adenosine triphosphate, one of the most widespread sources of energy for chemical processes occurring in the *cell*. The cell's supplies of ATP are constantly recharged by its *mitochondria* (Ch. 4).

Autonomic nervous system
The part of the nervous system largely outside voluntary control that regulates the smooth muscle in blood vessels, gut, bladder and elsewhere, and modulates the rate and force of the heart. It controls internal bodily functions like blood pressure, sweating and penile erection.

Axon
The slender process of the *neuron* that transmits onward signals away from the cell. A single axon leaves the neuron, although it will often branch as it nears its targets, allowing a single neuron to make numerous *synaptic* connections with other cells (Ch. 5).

Bacteria
Single-celled organisms, lacking compartments within their *cells*, related to the earliest life forms known to have existed on earth.

Basal ganglia
A collective term for an important group of *nuclei* (clusters of *neurons*) deep in the brain, including the caudate, putamen, globus pallidus and substantia nigra. In one major signalling loop involving the basal ganglia, signals pass from numerous areas of the *cortex* to the caudate/putamen (which function as a single unit), thence to the globus pallidus, on to the *thalamus*, and back to the originating region of cortex. Human diseases resulting from basal ganglia dysfunction include *Parkinson's disease* and *Huntington's disease.* While they are associated particularly with the control of movement they also contribute to the neural control of thought, personality and behaviour (Ch. 2).

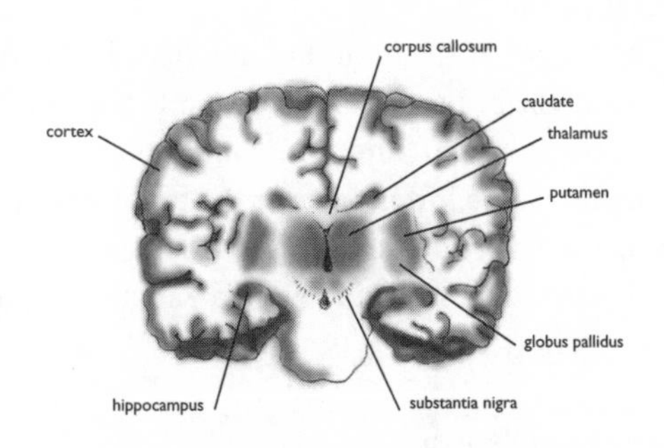

This figure looks into the centre of the brain, as it would appear if the front half were transparent. It picks out the interconnected regions of the 'basal ganglia', brain regions traditionally associated with the control of movement, but also involved in controlling thought, emotion and behaviour - the substantia nigra, where dopamine-producing cells are lost in Parkinson's disease, the caudate, putamen and globus pallidus. These are discussed in Chapter 2. It also shows the hippocampus, a key region in memory formation (Chapter 7), the corpus callosum, which allows exchange of information between the two halves of the brain (Chapter 8), the cerebral cortex (Chapter 8) and the thalamus, a major hub of interconnections between widespread brain regions.

Base – see **Nucleic acid**

Beta rhythm
Rapid rhythms, at 13–25 cycles/second, which can be recorded from the human scalp during mental activity.

Biochemical
A loose term I have used to refer to the chemical compounds that play an important part in living things like *DNA*, *proteins* and *carbohydrates.*

Biology
The science of life and living things. *Physics*, *chemistry* and biology are traditionally regarded as the three central areas of science. Scientists assume that results and theories in all three sciences will be compatible – that the facts of chemistry can be explained in terms of physics, and the facts of biology in terms of chemistry and physics. In practice it is doubtful that comprehensive reductions will ever be achieved.

Brain stem
The region of the brain linking the spinal cord below to the *cerebral hemispheres* above. The brain stem is divided into the *medulla* (closest to the spinal cord), *pons* (in the middle) and *midbrain* (closest to the hemispheres). Among other functions, the lower parts of the brain stem control breathing and the heart; the upper parts regulate the sleep–wake cycle by way of widespread connections to the hemispheres. In the United Kingdom 'brain death' is equated with death of the brain stem.

Brodmann's area
One of the areas of cerebral cortex distinguished by the neuroanatomist Korbinian Brodmann (1868–1918) using the light microscope. His areas differ in subtle anatomical features, such as the density of *cells* in particular layers of the *cortex*, but the boundaries between them have generally turned out to correspond to differences of function and his map therefore remains in current use (Ch. 8).

BSE (Bovine Spongiform Encephalopathy)
An infectious brain disease ('encephalopathy') giving rise to spongy ('spongiform') change in the brain in cattle ('bovine'). This condition is caused by an infectious protein (*prion*). Consumption of meat from cattle with BSE is thought to have been the cause of the human form of BSE, known as variant *CJD* (Ch. 3).

Calcium
An *element*, atomic number 20, familiar as one of the major ingredients of chalk, calcium carbonate. It is also abundant in the body, playing a key role in signalling processes, both in the course of neuronal firing, and within the cell. Entry of calcium into the *synapse*, for example, is required for the release of *neurotransmitters* from the terminals of *axons*. Entry of excessive amounts of calcium is a common terminal event in the lives of *cells* (Ch. 1).

Carbohydrates
A family of compounds of *carbon*, *oxygen* and *hydrogen* used to store energy and yield it to the *cell*; these include starch and cellulose in plants, and glucose and glycogen in animals.

Carbon
An *element*, atomic number 6, crucial to life on earth. *Atoms* of carbon form stable *molecules* with other atoms, including other atoms of carbon, creating the molecules on which life depends (Ch. 1). Organic chemistry is the *chemistry* of carbon compounds.

Cataplexy
Paralysis, complete or partial, on emotional arousal, particularly due to amusement, one of the cardinal symptoms of *narcolepsy* (Ch. 6).

Caudate nucleus – see ***basal ganglia***

Cell
Cells are, so to speak, the atoms of *biology*, the building blocks of complex forms of life, capable of leading an independent existence apart from their owner if they are carefully nurtured. Each cell contains the full complement of genetic material required to build a body, although in bodies like ours the cells in different *organs* use only the instructions relevant to their particular needs. The *neuron* is the principal specialist cell in the *nervous system*. There are around 100,000 million neurons in the brain (Ch. 5).

Cell membrane
The semi-fluid boundary of the cell at which it transacts its business with the outside world. It consists of two layers of (water-resistant) fat *molecules*, containing numerous *proteins* which communicate with other cells, for example

the protein *receptor* molecules which detect the presence of *neurotransmitters* released at *synapses* made by the *axons* of adjacent *neurons.*

Central nervous system

The brain and spinal cord (as opposed to the *peripheral* nerves which run to and from muscles and sense organs in the arms, legs, trunk and head).

Cerebellum

A region of the brain tucked in behind the *brain stem* and beneath the hemispheres, with a highly repetitive neuronal structure, conventionally associated with the smooth coordination of movement (but possibly also involved in the 'coordination' of thought and emotion). Signals pass from the cerebral *cortex* to the cerebellum, thence to the *thalamus* and back to the cerebral cortex in a control loop similar to the one involving the *basal ganglia.*

Cerebral hemispheres/cerebrum

The paired half-circles of the brain, covered by the folded cerebral *cortex*, each containing deep structures including the *basal ganglia* and the 'diencephalon', *thalamus* and *hypothalamus*, but not the *brain stem* or *cerebellum.*

Channel

The electrical activity of *neurons* is controlled by *proteins* situated in the *cell membrane* containing pores or channels which allow charged particles, like *sodium*, *calcium* and *potassium*, to pass in and out of the *cell.* Some channels, like those involved in transmitting the *action potential*, are opened by changes of voltage ('voltage-gated'); others are opened by the arrival of *neurotransmitters* ('ligand-gated'). A growing band of neurological disorders is turning out to be caused by defects in channels, including some types of *epilepsy* and migraine (Ch. 5) .

Chemistry

Like *physics*, chemistry studies matter and energy, but while physics focuses on their fundamental properties and behaviour, chemistry concerns itself with their more complex interactions. In particular chemists study how *atoms* of different *elements* combine in *molecules*, and how these molecules react together. Organic chemistry is the study of the extremely numerous molecules containing *carbon*, among them the fundamental molecules of life (like *DNA*, *proteins*, *fats* and *carbohydrates*).

Chlorine/chloride

Chlorine is a reactive *element*, atomic number 17, familiar to us as one half of the *molecule* present in table salt, sodium chloride. Like *sodium*, it was abundant in the ancient seas in which our ancestors evolved – and remains abundant in our body fluids, which preserve the composition of those seas (Ch. 1). When dissolved in water, atoms of chlorine often exist as electrically charged particles, chloride *ions*.

Chorea

Fidgety involuntary movements which result from disturbance of the function of the *basal ganglia*. 'Athetosis' describes related writhing movements. 'Choreoathetosis' combines the two types (Ch. 2).

Chromosome

The genetic material in each *cell* is condensed into 23 pairs of chromosomes, including the sex chromosomes (this pair comprises two X chromosomes in women, an X and a Y chromosome in men). Thus any given *gene* is located on a given chromosome, and will be present in each of us in two versions, one originating from our father, one from our mother (genes on the Y chromosome, and the X chromosome in men, are present in only one copy, Ch. 1).

Chronic fatigue syndrome

A state of disabling tiredness which does not, by definition, have any other simple medical or psychiatric explanation (Ch. 1).

CJD – see ***Creuztfeldt Jakob Disease***

Cognition

The sum total of our intellectual activity. The usual list of cognitive functions runs something like: attention, memory, executive function (ability to organise our thought and behaviour), language, praxis (our ability to perform skilled actions), perception (including spatial awareness). Consciousness may or may not be included in this list.

Coma

A state of impaired consciousness with no or much-reduced responsiveness associated with a variable degree of depression of brain activity. The eyes are closed. Coma is caused principally by conditions – like too much alcohol or too little oxygen – causing a widespead depression of brain function, or by more focal damage to the areas in the *brain stem* and *thalamus* which normally maintain wakefulness (Ch. 2).

Copper
A metallic *element*, atomic number 29, useful for making wire (and to describe hair of a corresponding colour) found in trace amounts in animal *cells*, with a key role in the functioning of some enzymes. The body's failure to handle copper normally lies at the root of Wilson's and Menke's diseases, discussed in Ch. 1.

Cortex
From the Latin for 'bark', the cortex is the folded outer surface of the *cerebral hemispheres*, their 'grey matter', rich in layered *neurons*.

Creuztfeldt Jakob Disease (CJD)
The most common human disorder caused by *prions*, occurring at a rate of one case/million population/year around the world. CJD is a rapidly progressive *dementia*, associated with signs of widespread disturbance in the brain's control of movement and sensation. Variant CJD (vCJD) is a novel form of CJD, occurring predominantly in the UK, caused by exposure to meat infected with *BSE* (Ch. 3).

Déjà vu
The disconcerting sense that our current experience echoes some ill-defined, elusive past experience. The phrase is also used colloquially to describe the experience of an event that is simply occurring for a second time (Ch. 7).

Delta rhythm
A slow electrical rhythm at less than 4 cycles/second that can be recorded from the scalp or brain during deep sleep or sometimes in *coma*.

Dementia
Literally, the loss of mind. Dementia involves the impairment of several cognitive abilities (for example of *memory* and language and perceptual abilities), often accompanied by changes in mood, personality and behaviour, with major impact on social functioning (Ch. 8).

Dendrite
The branching processes of *neurons* which receive the majority of signals from other neurons.

Diencephalon

The core of the *cerebral hemispheres*, comprising the *thalamus* and *hypothalamus.*

DNA

Deoxyribonucleic acid is the chemical material of our *genes*, packaged in our *chromosomes*: these transmit the instructions for building and maintaining our bodies – our genetic make-up (Ch. 2).

Dominance

The leading role of the left hemisphere of the brain in language has led to its being described as dominant (Ch. 8).

Dopamine

A *neurotransmitter* released in the *cerebral hemispheres* by *axons* originating in the *brain stem*, with a role in arousal, motivation and *motor* control. *Parkinson's disease* results from loss of dopamine-producing *neurons* in the brain stem. Schizophrenia is treated with drugs which block the action of dopamine in the *basal ganglia.*

Dualism

The philosophical tradition that draws a deep distinction between mental and physical events. Substance dualism distinguishes mental and physical substances; property dualism distinguishes mental and physical properties.

EEG

The electroencephalogram, a recording of the brain's electrical activity, usually recorded from the scalp.

Element

A pure substance containing *atoms* of only one kind: a diamond, for example, contains only atoms of the element *carbon*; a lump of pure iron only atoms of the element *iron.*

Encephalitis

An infection of the substance of the brain.

Encephalitis lethargica
A form of brain infection which became epidemic at around the close of the First World War. Disorders of arousal were prominent early in the course of the illness and in its long-drawn-out convalescence. It has all but disappeared.

Endorphin
One of the three families of opioid *neurotransmitters* important in pain signalling. The 'opiates', like opium and heroin, mimic their action in the brain.

Epilepsy
A disturbance of the brain's electrical activity causing abnormal synchronisation of activity, either locally ('partial epilepsy') or globally ('generalised epilepsy'). It can cause a wide range of disturbances of sensation, *memory* and thought as well as the more familiar 'grand mal' seizure.

Eukaryote
An organism with *cells* – like ours – that have internal compartments and *organelles* (Ch. 4).

Evolution
The process that has given rise to and shaped the diversity of living forms. The fundamental characteristics of a living thing are transmitted to its offspring in genetic material (now known to be *DNA*). Chance alterations (*mutations*) in this material cause variations in the characteristics of the offspring. Some of these will be advantageous, and enhance the individual's chances of reproduction. They will therefore spread through the population.

Fat
A second family of compounds of *carbon*, *oxygen* and *hydrogen* used to store energy and yield it to the *cell* (like *carbohydrates*) – but also to insulate us against heat loss, cushion our *organs* and sculpt the human form. Fat *molecules* are composed of glycerol, a sweet, sticky liquid in its pure form, combined with one, two or three 'fatty acids', most commonly palmitic, stearic and oleic (the last of these predominates in olive oil).

Fatal Familial Insomnia
A rare inherited form of *prion* disorder causing particularly prominent insomnia (Ch. 3).

Fibre
A term used to refer to the *axon.*

FMRI
Functional magnetic resonance imaging: one of the two main techniques which can be used to reveal the activity of the living brain as it performs specific tasks. The other is PET (see *functional imaging* below).

Frontal lobe
The foremost of the four lobes of the brain. It contains the *motor* cortex, and, broadly, is the lobe which governs the output of the brain, organising and regulating our behaviour (Ch. 8).

Fronto-temporal dementia
Dementia like Jan's (Ch. 8) that results from shrinkage of the brain's *frontal* and/or *temporal lobes* (see Ch. 8 for discussion of its sub-types).

Functional imaging
A group of techniques which makes it possible to visualise the brain regions activated by specific functions, ranging from seeing a flash of light to mathematical thinking. The two main techniques are PET (positron emission tomography) and *FMRI* (functional magnetic resonance imaging). The techniques rely on the fact that blood flow and energy consumption increase in active brain areas.

GABA
Gamma-amino butyric acid, the major inhibitory *neurotransmitter* in the *nervous system.*

Gene
Our genes are the inherited instructions for building our bodies, written in *DNA.* Each gene spells out the order of *amino acids* for a particular *protein molecule.* The Human Genome Project suggests that 30,000 genes are active in the human body (Ch. 2).

Global workspace
Bernard Baars's metaphor for the computational function of consciousness: it provides a central source of information that is widely available to specialised psychological subsystems.

Glutamate
A widespread excitatory *neurotransmitter*, released for example by *axons* running from *thalamus* to *cortex* and vice versa. It is an *amino acid.*

Glycine
An *amino acid neurotransmitter* which inhibits neuronal firing.

Golgi stain
The stain discovered by Camillo Golgi that highlights individual *neurons* in their entirety for reasons that are still uncertain. Ramón y Cajal used this stain to great effect in developing his theory that the *nervous system* consists of separate *cells* – neurons – communicating across a tiny gap – the *synapse* (Ch. 5).

Grey matter
Areas of the brain rich in *neurons*, such as the surface of the cerebral *cortex*, as opposed to the '*white matter*' which consists mainly of interconnecting fibre bundles.

Gyrus
One of the 'hills' of the folded cerebral *cortex*, as opposed to its valleys or '*sulci*'.

Hippocampus
From the Greek for sea-horse, the hippocampus is a curved structure tucked into the inner surface of the *temporal lobe.* It is required for the formation of new long-term conscious or 'declarative' *memories*, and to some extent for their retrieval as well (Ch. 7).

Histamine
An excitatory *neurotransmitter* with a role in maintaining wakefulness.

Hormone
A substance, often a small *protein*, that is released into the bloodstream and exerts a widespread effect on bodily functions – like insulin from the pancreas or adrenaline from the adrenal gland.

Huntington's disease
An inherited disorder affecting the brain, giving rise to fidgety movements – *chorea* – as well as disturbance of mood, thought, personality and behaviour (Ch. 2).

Hydrogen

The simplest of all *elements*, and the most ubiquitous in the universe, atomic number 1, a gas in its free form, bound to *oxygen* in water, H_2O (Ch. 1).

Hypocretin

A recently discovered *neurotransmitter*, manufactured in the *hypothalamus*, that helps to regulate the sleep–wake cycle. *Narcolepsy* is caused by deficiency of hypocretin (Ch. 6).

Hypothalamus

A small cluster of *cells* at the base of the brain lying beneath the thalamus, only the size of a small peanut, which controls the *autonomic nervous system*, regulating such things as appetite and thirst, but also sleep and waking. It is home to the *neurotransmitter*, *hypocretin*, that goes missing in *narcolepsy* (Ch. 6).

Hysteria

A condition in which a variety of seemingly neurological disorders – such as paralysis, loss of sensation and blackouts – occur in the absence of discernible neurological disease, often with some psychological factors in the background (Ch. 9).

Interneuron

A small *neuron* involved in short-range communication: a component of neuronal circuitry which intervenes between *sensory* neurons which bring information to and *motor* neurons which carry it away from the brain.

Ion

An *atom* or *molecule* carrying net electrical charge. Ions are important for neuronal signalling: the movement of ions, like those of *sodium*, potassium or *calcium* (which carry positive charge) or of chloride (which carries negative charge), into and out of neurons shapes the *action potential.*

Iron

A metallic *element*, atomic number 26, with a vital role in the body. It nestles within the oxygen-binding *protein*, haemoglobin, with which red blood cells are packed, and enables them to transport *oxygen* to *cells* throughout the body.

Kuru
A *prion* disease restricted to New Guinea, spread, mostly among women and children, by the habit of eating the brains of ancestors (Ch. 3).

Lead
A heavy metal, atomic number 82, prone to poison us, through the use of lead pipes, lead paints, leaded petrol and lead in industrial processes (Ch. 1).

Lesion
A very general term meaning a discrete area of damage or disease, caused by nature or by man (an 'experimental lesion').

Lewy Body Dementia (LBD)
The third most common variety of *dementia*, characterised by the prominent occurrence of hallucinations, marked (hour to hour or day to day) fluctuations in cognitive function and features of *Parkinson's disease* (tremor, stiffness, paucity and slowness of movement, unsteadiness). 'Lewy bodies', abnormal inclusions found within neurons in a region of the *brain stem*, are the pathological hallmark of Parkinson's disease. In LBD they are found in the cerebral *cortex* (Ch. 8).

Limbic system
A network of structures at the 'limbus' or border of the brain, with a key role in emotion and *memory*. It comprises the *hippocampus* and surrounding *cortex* in the medial *temporal lobe*, parts of the *thalamus*, and other linked areas including the cingulate gyrus in the medial *frontal lobe* (i.e. its inward-facing part) and the amygdalae (Ch. 7).

Lobe
One of the four major subdivisions of the *cerebral hemispheres* – *frontal, parietal, occipital, temporal* (see Ch. 8).

McLeod's syndrome
An inherited disorder of movement, thought and behaviour, due to mutations of a *gene* that travels on the X *chromosome*, the XK gene (see Ch. 2).

Magnesium
A reactive *element*, atomic number 12, present in our body fluids.

Medulla
The part of the *brain stem* closest to the spinal cord, crucial for life because of its role in maintaining breathing and the activity of the heart.

MELAS
A neurological disorder caused by a mutation in the *DNA* contained within our *mitochondria*, and passed from mother to child (as all the DNA within our mitochondria is inherited from our mothers). It stands for '*myopathy*, encephalopathy, lactic acidosis and stroke-like episodes': see Ch. 4, 'The Children of Eve', pp. 72–4, for further explanation.

Membrane – see ***cell membrane***

Memory
Broadly, this is the capacity that allows our experience and behaviour to change as a result of what has happened to us in the past. Memory can be classified in many ways, but one especially useful subdivision – because it corresponds to a clear-cut biological distinction in the brain – is into declarative and procedural types. Declarative memories can be articulated. They are further subdivided into episodic memories, for one-off events, and semantic memories, for our database of knowledge about language and the world (what a table is, who is the President of the USA . . .). Procedural memories are ones that we demonstrate we possess by doing something differently – for example riding off on a bike (Chs 7, 8).

Midbrain
The part of the *brain stem* closest to the brain proper. Damage here is particularly likely to impair consciousness.

Millisecond
One thousandth of a second.

Mitochondrion
One of the minute chemical machines, or *organelles*, within the *cell*. Mitochondria are required for energy generation within the cell. Unlike the other organelles, they contain their own *DNA*. They are thought to have originated as independent organisms which entered into a symbiotic relationship with the ancestor of our cells about 2,000 years ago.

Molecule
Two or more *atoms* linked by chemical bonds. Water is a simple molecule (H_2O, two atoms of hydrogen coupled to one atom of oxgen); *proteins* are complex ones.

Motor
To do with movement. So the 'motor cortex' is the part of the *cortex* which most directly influences our choice of movement.

Mutation
A change in a *gene* that can give rise to a change in a *protein* – for better or for worse. If the change is for the better, the mutation may be selected and spread through the population. Mutations are the raw material for *evolution* (Ch. 2).

Myalgic encephalomyelitis (ME)
An alternative term for *Chronic fatigue syndrome*, which is probably falling out of favour. It means, literally, 'muscle aching inflammation of brain and spinal cord', which is puzzling, as any clear-cut evidence of inflammation of brain and spinal cord would point to some other better-characterised neurological disorders and rule out a diagnosis of ME.

Myelin
The fatty insulating sheath that is wrapped around many *axons* in the *nervous system* to increase the speed with which they can conduct signals.

Myopathy
A disorder ('pathy') of muscle ('myo') (Ch. 1).

Narcolepsy
A sleep disorder characterised by excessive daytime sleepiness, usually accompanied by *cataplexy* (paralysis on emotional arousal, especially during laughter), and often also by sleep paralysis (inability to move on first awakening from a dream), hypnagogic hallucinations (dream-like imagery on closing the eyes to sleep), and disturbed nocturnal sleep. It is caused by loss of the recently discovered *neurotransmitter hypocretin* (Ch. 6).

Nerve
The collective term for a group of *neurons* whose fibres run together. As a rule some will be supplying muscle, others returning from sense *organs* (although some nerves, like the optic nerve, are purely *sensory*).

Nervous system
The brain, spinal cord and peripheral nerves: all the nervous tissues of the body.

Neural correlate of consciousness (NCC)
This term is used to refer to the neural basis of conscious experience: the combination of structure (the 'where' of consciousness) and *physiology* (the 'how' of consciousness) that collectively give rise to our experience (or so one popular view of consciousness holds).

Neural network
Natural neural networks are interconnected groups of *neurons* that collectively undertake a particular function, like the network of neurons linked in the l*imbic system* that play a crucial role in *memory* formation.

Neurology
The branch of medicine which deals with disorders of the *nervous system* (closely allied to neuroscience, the scientific study of the nervous system).

Neuron
A *nerve cell* (Ch. 5).

Neurotransmitter
A chemical released by a *neuron* at its junction (or *synapse*) with another *cell*, which generally has the effect of increasing or reducing the level of excitation in the second cell, thereby increasing or decreasing the rate at which it transmits signals. Neurotransmitters can also exert more subtle effects by way of *second messenger* systems (see Ch. 6, section v).

Nitrogen
A gaseous *element*, atomic number 7, one of the ubiquitous elements in living things in which *atoms* of the gas are 'fixed'. Nitrogen is particularly abundant in *amino acids*, the building blocks of *proteins*.

Noradrenaline
This *neurotransmitter* is called norepinephrine in the USA. Closely related to adrenaline, a hormone released by the adrenal gland to help mobilise the body's resources for 'fight or flight', noradrenaline is one of the neurotransmitters of the brain's 'activating system'. It is synthesised in the *brain stem* and released widely through the brain.

Nucleic acid

DNA (deoxyribonucleic acid) and RNA (ribonucleic acid) are 'nucleic acids', giant *molecules* composed of chains of sugar molecules (deoxyribose in DNA, ribose in RNA), linked with one another by phosphoric acid (phosphate) molecules. Each sugar molecule is also linked to a nucleotide base (adenine, guanine, cytosine, thymine in DNA, with substitution of uracil for thymine on RNA), giving rise to the following kind of structure:

sugar-base
/
phosphate
/
sugar-base
/
phosphate . . .

The ordering of bases provides the basis of the genetic code, 'cytosine–adenine–guanine', CAG, for example specifying the *amino acid* glutamine. In the double helix of DNA (Ch. 2), two long nucleic acid molecules twist around another, the bases between them forming loose chemical bonds (adenine always pairing with thymine, guanine with cytosine):

sugar-adenine . . .	. . . thymine-sugar
/	/
phosphate	phosphate
/	/
sugar-guanine . . .	. . . cytosine-sugar
/	/
phosphate	phosphate

Nucleotide base – see ***nucleic acid***

Nucleus

A term with two entirely separate biological meanings. In cell *biology*, the nucleus is the part of the *cell* which contains *DNA*, the headquarters of chemical operations. In *neurology*, a 'nucleus' is a cluster of *neurons* which generally share a function or functions.

Occipital lobe
The hindmost of the four lobes of the brain. It contains the primary visual *cortex*, right at the back of the brain (the 'occipital pole'), and several of the other visual areas (Ch. 8).

Opioids
A family of *peptide neurotransmitters* with a major role in modulating pain. Their existence in the undoctored brain explains the effects of the opiates, like heroin, which are both prescribed and abused.

Organ
This word is used to refer to the body's major internal parts, like kidney, liver, heart – and the hero of this book, the brain.

Organelles
The residents of the interior of our *cells*, including the *nucleus* and *mitochondria.*

Organism
A living thing.

Oxygen
An *element*, atomic number 8, essential for much life on earth, allowing efficient energy production from our foodstuffs, via a carefully regulated 'burn' (Ch. 1).

Parietal lobe
The lobe of the brain bordered by the *occipital* behind, the *temporal* below and the *frontal* ahead. It contains the 'somatosensory *cortex*' which receives *sensory* information about touch and joint position, and plays a major role in our appreciation of spatial relationships (Ch. 8).

Parkinson's disease (PD)
A common neurological disorder, caused by lack of the *neurotransmitter dopamine* in the brain, leading to any combination of tremor, stiffness, paucity and slowness of movement, and unsteadiness.

Peptide
A short string of *amino acids.*

Periodic table
A table displaying all the known *elements*, organised in a manner that highlights the patterns apparent in their chemical behaviour (see Ch. 1).

Peripheral nervous system
The nerves beyond the brain and spinal cord: the cranial nerves in the head, numbered 1–12, and the numerous nerves in the limbs and trunk which run out to muscle and back from sense *organs* like those in skin and joint.

Phosphorus
An important *element* in living things, atomic number 15. It contributes for example to the *molecule* that stores energy for immediate use within the *cell*, synthesised through the oxygen-demanding work of *mitochondria*, *ATP* or adenosine triphosphate.

Phylogeny
The development of the species, as opposed to ontogeny, the development of the individual.

Physics
The most fundamental of the sciences, the study of matter and energy.

Physiology
The study of how the body works, the functioning of its parts.

Plasticity
The capacity of the *nervous system* to adapt to change. Neural plasticity, the basis of learning and *memory*, depends mainly on the modifiability of *synapses*, the connections between *neurons*: these can multiply, wither and alter their strength according to the play of signals across them (Ch. 7).

Pons
The central part of the *brain stem*, lying above the *medulla* and below the *midbrain* ('pons' from Latin for bridge, as the front view of the pons in the intact brain looks like a bridge between the two halves of the *cerebellum*).

Potassium
An element, atomic number 19. A soft metal, it is present in relatively high concentrations dissolved in the fluid within *cells*, where it exists in a positively

charged form as a potassium '*ion*'. It plays a key role in neuronal firing, flowing out of *neurons* during the later part of the *action potential*.

Prefrontal cortex
The region of the *frontal lobe* that lies beyond the *motor* and premotor *cortex*. Activity in motor and premotor cortex is more or less tightly linked to the control of movement; the prefrontal cortex exerts a less direct influence on behaviour, in decision-making for example (Ch. 8).

Prion
Prion *protein* is a normal constituent of the brain, but in disorders like *CJD* prion protein *molecules* undergo a chemical change that renders them both indigestible by the *cells*, and confers on them the ability to convert *molecules* of normal protein into the indigestible form (Ch. 3).

Projection
A bundle of *axons* conveying signals from one part of the *nervous system* to another: for example, there are several important projections from the *brain stem* to the hemispheres which influence our state of arousal.

Prokaryote
An organism, like a bacterium, with *cells* that lack internal compartments and *organelles* (Ch. 4).

Prosopagnosia
Loss of the ability to recognise faces (despite otherwise well-preserved visual abilities).

Protein
A *molecule* formed by a sequence of *amino acids*, specified by a *gene*, from the Greek for 'first things'. Proteins build, maintain, support and run the human body (under supervision from the genes which are a step further removed from the immediate business of life: Ch. 3).

Psychiatry
The branch of medicine concerned with mental disorders. Psychiatrists – unlike *psychologists* – are medically qualified, but work closely with psychologists in assessing and treating disorders of the mind. The overlap between *neurology* and

psychiatry is so great that they are arguably subdivisions of a single medical discipline.

Psychology
The science of mind: a hybrid subject, psychology embraces first-person approaches to the study of mind – direct reports of experience – as well as third-person approaches, such as the investigation of processes in the brain that make experience possible. It therefore bridges the humanities, which place human experience at centre stage, and the sciences, which investigate the world (so far as is possible) as it exists independently of our experience. Clinical psychologists have an academic background in psychology, but work with patients, either in a diagnostic role (assessing *memory* problems, for example) or as therapists (for example treating phobias or depression using any of several 'talking therapies').

Psychosis
The group of psychiatric disorders characterised by the occurrence of delusions (fixed false beliefs) and hallucinations.

Pyramidal cells
Large cortical *neurons* with a pyramidal shape. Their *axons* leave from the base of the pyramid; their extensive dendritic trees include a major branch from the apex. Most pyramidal cells send their axons out of the immediate vicinity, into the *white matter*: they are therefore the '*projection*' or output cells of the cerebral *cortex*.

Recall
The ability to retrieve a *memory*, with or without a cue.

Receptor
A *molecule* adapted to receive and lock on to another, with some resulting effect, like the opening of a channel allowing the flow of particles into or out of the *cell*. In neuroscience, receptors are usually on the lookout for *neurotransmitters*, for example the *acetylcholine* receptor in muscle which picks up the neurotransmitter released by a motor neuron – acetylcholine – causing the muscle to contract: Ch. 6).

Recognition
Knowledge that something or someone is familiar: we sometimes recognise things and people without being able to *recall* anything about them.

Reductionism
The philosophical view that talk about a complex or mysterious function or entity, like a conscious state, can be reduced to talk about some simpler or better understood function or state, like activity in the brain.

Reflex
A more or less automatic fragment of behaviour, like the jerking of the knee when it is tapped with a tendon hammer. Even reflexes like this one are not completely automatic and unalterable (Ch. 7).

Reflex epilepsy
Epilepsy regularly provoked by a particular trigger: a flashing or flickering light is a common trigger, but a wide range of activities including reading, calculating and summoning up a particular *memory* have been described as triggers.

REM
Rapid eye movement: one of the characteristic features of dreaming sleep.

Reticular formation
From the Latin *reticularis*, meaning net, the reticular formation is the region running the length of the *brain stem* that regulates functions like breathing and the beating of the heart in its lower reaches, wakefulness and arousal in its upper parts.

Scrapie
A *prion* disease affecting sheep, thought, probably, to be the source of the related disease affecting cattle, bovine spongiform encephalopathy (*BSE*: Ch. 3).

Second messenger
A chemical generated by the activation of the family of receptors which respond to the arrival of a *neurotransmitter* by further chemical signalling within the *cell*, rather than by directly opening or closing an *ion* channel (Ch. 6).

Seizure
A general term for an abrupt disturbance of experience or behaviour, often due to *epilepsy*.

Sensory
To do with sensation: the optic *nerve*, for example, is a sensory nerve, conveying information about visual sensation from eye to brain.

Serotonin
A *neurotransmitter* originating mainly in the *brain stem* with an important role in regulating arousal, appetite and mood.

Sleep stages
Sleep has an internal structure: it is usually divided into four deepening stages of non-rapid eye movement sleep (the two deepest comprise *slow wave sleep*) and a further stage of 'paradoxical' or 'rapid eye movement' sleep, in which we dream.

Slow wave sleep
Sleep characterised electrically by a predominance of 'slow waves' (*theta* and *delta rhythms*). Deep slow wave sleep predominates in the early parts of the night, with increasing amounts of lighter slow wave and REM sleep as the night proceeds.

Sodium
Sodium is a reactive *element*, a metal, atomic number 11, familiar to us as one half of the *molecule* present in table salt, sodium chloride. It was abundant in the ancient seas in which our ancestors evolved – and remains abundant in our body fluids, which preserve the composition of those seas. Its *atoms* often exist in an electrically charged form, as positively charged sodium '*ions*': rapid movement of sodium ions into the *neuron* – where its concentration is kept relatively low – is the key event in the 'firing' of the neuron when it is sufficiently excited – its *action potential* (Chs 1, 5).

Stimulus
A term now used rather loosely to refer to the item used as the target or trigger in an experiment or of a given type of behaviour.

Stroke
A general term for an episode of neurological impairment caused by death of a part of the brain due to trouble with a blood vessel (usually blockage of an artery). Stroke is very common, and much of our established knowledge of the functions of the human brain has come from studies of its effects.

Subcortical
Term used to refer to structures and processes which occur below the *cortex*. For example, the *thalamus* is an important subcortical *nucleus*.

Sulcus

One of the depths or valleys of the folded cerebral *cortex*, as opposed to its summits or '*gyri*'.

Synapse

The point at which one *neuron* makes contact with another, usually chemical contact. This involves the release of chemical *neurotransmitter* by the presynaptic ('before-synaptic') *cell* which travels across to receptors on the surface of the post-synaptic ('after-synaptic') cell (Ch. 6).

Synchrony

The firing of *neurons* in time with one another.

Syncope

From the Greek *syncoptein*, to cut, this term describes temporary loss of consciousness due to interruption of the blood supply to the brain, as in a faint.

Temporal lobe

The lowest-lying lobe of the brain, above the ear, with a special importance for hearing, smell, visual recognition, language comprehension and *memory* (Ch. 8).

Thalamus

A large *nucleus* lying at the centre of the *cerebral hemispheres* around the sides of the third *ventricle* (see sketch under entry for **basal ganglia**). The thalamus receives much of the *sensory* information (relating to vision, hearing and touch) that eventually reaches the cerebral *cortex*, much of the activating input from the *brain stem* destined for the cortex, and transmits signals which loop between the cortex and the *basal ganglia* and the cortex and the *cerebellum*. It is therefore a microcosm of cortical activity, and damage here is keenly felt by the rest of the brain.

Theta rhythm

A slow electrical rhythm at 4–8 cycles/second which can be recorded from the scalp or brain during sleep or sometimes in *coma*.

Ventricle

One of the fluid-filled spaces at the centre of the brain. These spaces enlarge under pressure in 'hydrocephalus' (water on the brain).

Virus

Packets of *genes* wrapped up in *protein*, viruses lie on the borderline of life: they hijack the machinery of the *cells* they enter to reproduce and propagate themselves, usually destroying them in the process.

Visual cortex

A general term referring to the parts of the cortex concerned with vision, including the primary visual cortex but also the thirty or so further visual areas with more or less specialised roles in visual processing.

White matter

The bundles of fibres interconnecting areas of brain are white because of the concentration of insulating *myelin* wrapped around *axons*. White matter contrasts with the '*grey matter*' of the cerebral *cortex* and deep nuclei of the brain, which contain aggregations of neuronal *cell* bodies.

Appendix

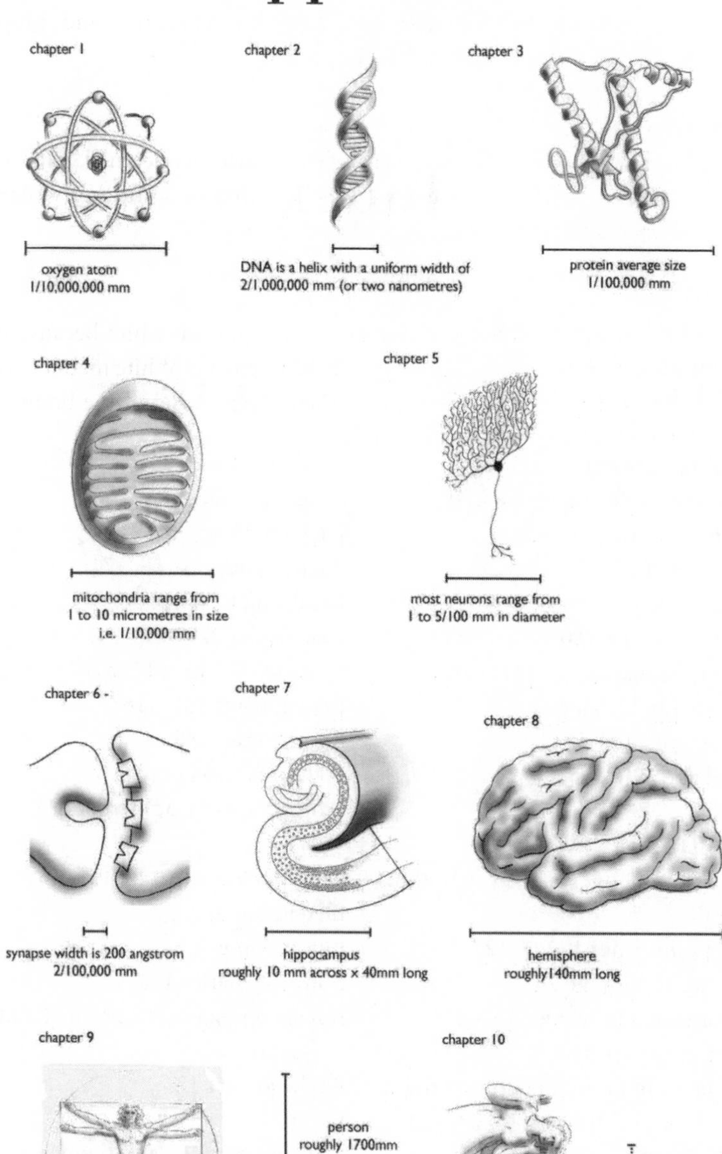

chapter 1
oxygen atom
1/10,000,000 mm
chapter 2
DNA is a helix with a uniform width of
2/1,000,000 mm (or two nanometres)
chapter 3
protein average size
1/100,000 mm
chapter 4
mitochondria range from
1 to 10 micrometres in size
i.e. 1/10,000 mm
chapter 5
most neurons range from
1 to 5/100 mm in diameter
chapter 6 -
synapse width is 200 angstrom
2/100,000 mm
chapter 7
hippocampus
roughly 10 mm across x 40mm long
chapter 8
hemisphere
roughly140mm long
chapter 9
person
roughly 1700mm
chapter 10
soul

Index

'g' indicates a glossary entry, (f) indicates a relevant figure.